WORLD WAR 1
Trench Warfare

WORLD WAR 1
Trench Warfare

Written and compiled by Michael Houlihan

SUPER SOURCE BOOKS   Ward Lock Limited · London

© Ward Lock 1974

ISBN 0 7063 1826 9

First published in Great Britain 1974 by Ward
Lock Limited, 116 Baker Street, London,
W1M 2BB

Designed by Conal Buck

Text filmset in 10pt Monophoto Times
by Keyspools Ltd, Golborne, Lancs.

Printed and bound in Great Britain by
Tinling (1973) Ltd

*All the photographs displayed in the text are
reproduced by kind permission of the
Imperial War Museum.*

*The maps and diagrams were prepared and
executed by Ray Allen.*

Preface

This book does not attempt to catalogue and recount the main military and political events which occurred on the Western Front from August 1914 until November 1918. Instead, attention has been focused upon the tactical factors which contributed to the creation and maintenance of the defensive deadlock. Inevitably, limitations of space do not permit a profound examination of these influences. The text, therefore, must serve as only a simple guide to the more obvious features and as a link between the photographs. However, it is hoped that the photographs themselves will convey a description of trench warfare which cannot be matched by words.

Michael Houlihan

1   The Pre-War Origins Of Trench Warfare

The Growth of Mass Armies

The strategic and tactical origins of trench warfare can be traced to the economic and technological forces liberated by the Industrial Revolution. By the beginning of the twentieth century, the political and military leaders of the foremost Powers in Europe found an unprecedented stock of technical knowledge, sophisticated *matériel* and manpower at their disposal. Militarily, this meant that, in an age of growing nationalism, the individual citizen could be more closely identified with his personal share in the defence of the state by means of conscription. Similarly, these economic assets added a new dimension to the influential theories of Karl von Clausewitz, the Prussian military philosopher. His book *On War* was published posthumously in 1832, and had

moulded the thinking of several generations of military leaders. Basically, Clausewitz advocated mass warfare and the importance of concentrating superior forces against the enemy's main force. However the military leaders of 1914 misinterpreted his tenets by neglecting modern technological and organizational realities. The belief in a war of rapid mobility and short duration ignored the economic and industrial systems which could maintain a mass army in the field, almost indefinitely, and equip it with the most advanced weapon-power.

The prevailing continental confidence in the principle of the mass army had largely been created by the victories of the Prussian conscript armies against the Austrians and French in 1866 and

British trenches before Sebastopol during the Crimean War. Sixty years later mobile warfare was to degenerate into similar siege conditions, only on a much vaster scale. (IWM Neg Q71084)

7

1870 respectively. It was believed that a large, strong army provided national security and a deterrent against aggression from external Powers. However, by its very nature, the mass army contributed to the accelerating spiral of fear which caused the countries of Europe to seek further security by a complicated system of alliances. This delicate balance between rival power blocs could be upset by even a minor clash of conflicting interests. This critical situation was aggravated further by the military plans of the Alliances. These relied upon the ability to strike the enemy first in a rapidly developed and mobile campaign. An essential element, therefore, in each nation's war plan was the time-table of mobilization.

Railways: Strategic Mobility

The problem of rapidly and efficiently concentrating and deploying a mass army in the field was one which each General Staff had been obliged to work out a long time before war was declared. By 1914, the solution of this logistical problem hinged directly upon the effectiveness of a nation's railway system. Refusing to rely upon the vagaries of the commercial railway companies, each General Staff maintained a Railway Section which was responsible for the planning and revision of the rail movement timetables. They also advised where new lines should be laid. For example, Germany increased the number of double lines to the Western Frontier from nine to thirteen between 1870 and 1914, mostly for military purposes. The Railway Sections also supervized the design of frontier terminals; some village platforms on the German frontier with Belgium were half a mile long. On the threat of war, the General Staffs were fully geared to take over the national railway network in each country. For all combatants, the quantities of men and materials to be despatched were enormous. The German plan called for the movement of 20,000 trains in seventeen days. The British Expeditionary Force was required to move to an English railway timetable as well as combine its movements with those of the French once the troops and

Federal entrenchments at the foot of Kenesaw Mountain in 1864. The American Civil War realistically illustrated the superiority of defensive firepower over the frontal assault. (IWM Neg Q45179)

matériel had crossed the Channel.

When the troops arrived at the railhead, they had to cover the final distances into the battle area on foot. This tactical sluggishness of large numbers of men on the battlefield, sacrificed the strategic advantages of mobility and concentration conferred by the railway. Throughout the war, the lack of tactical mechanization in the attack contrasted with nimble defensive techniques which could rely on reinforcements being brought up by rail before the enemy could break through and exploit the

situation on foot. Mechanization in warfare was not only a question of self-propelled guns and tanks; it also involved the efficient transport of men and supplies as well as the servicing of field units with satisfactory communications. Yet, in moulding their tactical doctrines, both sides neglected the advances which had been made in the electrical, chemical and automobile industries during the last quarter of the nineteenth century.

The Developement of Firepower

The failure to solve the tactical problem of mobility directly contributed to the dominance of defensive firepower in the field. Furthermore, improved industrial processes had brought about a radical change in the tactical influence of small arms. By the end of the nineteenth century, highly effective and well-finished weapons could be produced relatively cheaply and made available in bulk. The basic improvements consisted of the widespread introduction of rifling, breech-loading and the combined bullet and charge or cartidge. These features

provided the essential foundation for the introduction of subsequent developments such as the bolt action, the metallic cartridge and rimless round, smokeless powder and the box magazine. By 1914, the ordinary conscript soldier was equipped with a weapon which combined considerable accuracy with rapidity of fire and an unprecedented stopping-power.

A further significant, yet largely unrecognized, development was the introduction of the machine-gun into military service. The simple firing system, evolved by Hiram Maxim, was adopted by both the British and German armies. The Maxim machine-gun used the power of the recoil forces, generated by the explosion of the powder charge, to produce the entire cycle of operation. This included the whole sequence of feeding in cartridges and ejecting the empty case. The weapon was capable of 600 rounds per minute. It was introduced into British service in 1891 and was used in the Matabele Wars (1893–4), on the North-West Frontier of India (1895) and at Omdurman (1896) with remarkable success. A larger version,

Fredericksburg, Virginia, during the American Civil War. The scene of much bitter fighting, Fredericksburg reflected the importance of logistical targets. (IWM Neg Q45173)

11

British infantry during the Boer War. As a result of lessons learnt during this war, British infantry were the best trained troops in Europe by 1914. (IWM Neg Q71941)

the 37-mm 'pom-pom', was used by the Boers against the British during the South African War (1899–1902). By 1914, most armies were equipped with a ratio of two machine-guns to every thousand men. However, even this limited number was capable of crippling the attack of a thousand troops.

The introduction of the machine-gun and of rifled and breech-loading artillery and infantry weapons had swung the tactical balance in favour of the defensive. The stopping-power of modern fire-arms and artillery had been amply illustrated during the American Civil War (1861–5) and the Russo-Japanese War (1904–5). However most of the military leaders of Europe disregarded the lessons and signs which contradicted their belief in the superiority of the offensive.

2   The Opposing Armies

The British Army in 1914

The Crimean War (1854–6) represented the last occasion when British troops had fought alongside European troops in a major continental conflict. Since 1856, the small volunteer British Army had been employed in numerous minor colonial campaigns against badly equipped and ill-organized enemies. It was not until the South African War (1899–1902), fought against highly mobile, well armed and resourceful marksmen, that the inadequacies of the British military system were ruthlessly exposed. In 1903, the Report of His Majesty's Commissioners on the War in South Africa indicted every facet of the Army from its organization and administration to its equipment and training.

The defects in Britain's home defences had also become apparent during the Boer War. In 1899, out of a higher establishment strength of 249,460 regular troops and 9,000 reservists, 130,000 men were required for the defence of the Empire, leaving only 70,000 for overseas service. Although these latter troops also formed an integral part of the home defence system, they had to be employed against the Boers. In the past, Britain had relied upon the Royal Navy for protection, but the growing German naval menace and the Kaiser's overt support for the Boers had caused Britain to abandon her policy of isolation from Europe and to re-examine the whole question of defence.

In 1902, the Committee of Imperial Defence was established and made responsible for the co-ordination of Imperial strategy. Two years later, the Report of the War Office (Reconstitution) Committee, better known as the Esher Report, established the system by which military administration was placed in the hands of the War Office, leaving Field Officers freer to concentrate on their military duties. An Army Council was also created to direct military policy and a revised General Staff now co-ordinated its implementation. The new post of Chief of the Imperial General Staff replaced that of Commander-in-Chief which was abolished.

The recommendations of the Esher Report were given substance largely by R. B. Haldane, who became the Secretary of State for War at the end of 1905. He was responsible for the reorganization of the home field army and the reserve system. Haldane's reforms were based upon the need to create an expeditionary force which could be rapidly mobilized and deployed in Europe. Britain's increasing diplomatic involvement in Europe was illustrated by the unofficial and, later, official contacts between the French and British General Staffs, from 1905 to 1914. A provisional plan was outlined to place a British expeditionary force of six infantry divisions and a cavalry division, a total of 120,000 men, on the left flank of the French Army in the event of war with Germany. In spite of this, Britain did not abandon her policy of maintaining a small, professional volunteer army. The Royal Navy still remained the first line of defence and, with the Liberal Party's initiation of a programme of social reform, political and financial factors prevented the introduction of conscription, even if the British public had been prepared to forget their traditional hostility to the concept of a large standing army.

The British Expeditionary Force was created from regular troops available under the two-battalion regimental system. This avoided having to increase the size and cost of the Regular Army. The B.E.F. was undoubtedly the best trained, best equipped and best organized force Britain had ever possessed, and its standard of musketry became unique among European armies. However its small size meant that it was hardly geared to warfare on a continental scale. To provide a support system and basis for expansion, Haldane created the Territorial Force out of the old militia and volunteer formations in 1908. Although it was recognized that the members of the non-professional Territorial Force could volunteer for service abroad, if the need arose, it was primarily intended for home defence. It was organized into fourteen infantry divisions and fourteen cavalry brigades and was administered by County Associations under Lords-Lieutenant. A Special Reserve was also created to garrison Regular Army depots and replace losses in time of war. The countries of the Empire, apart from India, did not maintain regular armies, only small defence and garrison forces.

The reorganization of the British Army was backed up by a comprehensive reform of training, based on revised tactical manuals and an improved staff system. However some of the tactical lessons of the Boer War were deceptive. Although the effects of modern firepower had been correctly gauged, the mobility and rapid manoeuvre seemed to confirm the traditional belief in the decisive tactical role of cavalry. Thus the conventional approach of many senior general officers to modern military problems nullified the advanced fieldcraft techniques of the regular soldier.

The German Army in 1914

The organization of the German Army in 1914 was established upon the Prussian concept of the 'Nation in Arms' or the involvement of every able-bodied male citizen in the defence of the State. The basic theory was that an effective army could be moulded from short-service conscripts, but on the condition that a highly-trained nucleus of long-service Officers and NCO's provided the tactical leadership. The theory was amply justified in 1870 when the Prussian Army, under the outstanding leadership of von Moltke the elder, was largely responsible for the defeat of the long-service French Army.

The German system of universal military service was imitated by several European armies, but none were able to match the efficiency of the German organization. Most German males became liable for military duty at the age of seventeen, when they registered for

service. At twenty the individual began his service proper, usually spending a total of seven years with the colours and in the Reserve. In the foot artillery and infantry, two years were spent with the colours and five in the Reserve; in the horse artillery and cavalry, three years were passed with the colours. The Reserve was intended to bring the field army up to war establishment and to augment the number of active units available, as soon as mobilization began. Service in the Reserve required the conscript to attend two training periods, neither of which lasted more than eight weeks. Reserve duty was followed by five years in the First Ban of the *Landwehr* and then service in the Second Ban until the age of thirty-nine. The purpose of the *Landwehr* was to form extra units of cavalry and infantry or, in the case of other branches of the service, to supplement shortages in the field army. Following *Landwehr* service, the conscript passed into the *Landsturm*, until the age of forty-five when he was released from military obligations. The *Landsturm* was relied upon to supply depot troops on the outbreak of

A German cavalry patrol. In 1914 cavalry represented the only force, on both sides, capable of rapid mobility. (IWM Neg Q42054)

war and was not liable for training or muster during peacetime. A further source of depot troops was the *Ersatz* Reserve, a body largely constituted of men exempted from the normal active service duty, but who were liable for periodic training over twelve years.

The overall organization of the Imperial Army was based upon the territorial division of all the German States into twenty-four military districts. Each district, focused upon a major town or city, formed the recruiting and garrison area for an Army Corps of two divisions. The eight infantry and four artillery regiments which comprised the Corps were each allocated a permanent station in a subsidiary town, as was the separately-organized cavalry regiment. At mobilization, the individual units of the Corps concentrated at a pre-arranged assembly point. Since each active Corps was duplicated by a Reserve Corps, the military district provided seventy-four infantry battalions of varied quality, six artillery and three cavalry regiments, in addition to auxiliary troops, on the outbreak of hostilities. The Reserve Corps was raised alongside the active Corps but existed as an individual tactical unit, led by regular officers and organized and trained to front-line standards. This unprecedented use of Reserve alongside active Corps gave the German Army a tactical superiority in the concentration of numbers which was to dislocate French offensive war plans in 1914.

The German General Staff was composed of well-trained professional officers, specially selected for their administrative ability and their attention to organizational detail. However too much repetitive practice, geared to a single plan, and inflexibility of approach tended to blunt originality and suppleness of reaction to uncalculated situations. In the ranks, a thorough training in mass tactics failed to develop individual initiative, a problem common to all conscript armies in 1914.

The German Army in 1914 was probably the best equipped in Europe to cope with the problems which were to be posed by trench warfare. However the pre-war scale of preparation for siege conditions was not sufficient for this initial advantage in howitzers, mortars, machine-gun deployment and personal weaponry such as grenades, to be effective.

The French Army in 1914

The decisive defeat of the long-service French Army in the Franco-Prussian War, led to a radical re-shaping of French military organization and thought. France, haunted by the steady expansion of German military strength, decided to create a mass conscript army. Believing that the disaster of Sedan was due to a defensive attitude, French military theorists advocated a return to the tactical traditions of the Napoleonic era and the concept of the superiority of the offensive under all circumstances.

The fundamental French problem was to construct a new mass army, in peacetime, which could equal, man for man, the yearly intake of the mature German conscript system. The principle of universal military service was reintroduced in the military reforms of July 1872 but, although the former system of 'substitution' was abolished, too many exemptions were still allowed, particularly in favour of the wealthier classes. In 1889, the terms of service were broadened to make military duty compulsory for all able-bodied men. Three years were spent with the colours, seven in the Reserve, a further six in the Territorial Army and nine years in the Territorial Reserve. It was hoped that this measure would eventually raise the total number of trained men available on the outbreak of war from 2,000,000 to 3,000,000.

After 1890, changes in the balance of population between France and Germany revealed further defects in the size of the French Army. In an attempt to redress the growing disparity, more conscription laws were passed in March 1905 and October 1913. Under the terms of the latter law, conscripts were obliged to serve three years in the field army and a further seven in the Territorial Army and a further seven in the Territorial Reserve. It was expected that this new system would provide a larger Reserve and more thorough training, as well as giving a greater measure of security during the critical first days of mobilization. However the outbreak of war came before the provisions of 1913 had taken full effect.

The organization and internal administration of the French Army was based on twenty-one Army Corps districts, including the XIX Corps in North Africa. This provided forty-seven divisions, including white and colonial units. The infantry, cavalry, artillery and engineers, as well as the *Chasseur* or rifle battalions were recruited throughout the Army Corps district so that, on the declaration of a state of war, the corps mobilized in its own territorial

French observers during the Battle of the Aisne in September 1914. Tactically and sartorially the French were unprepared for modern warfare. (IWM Neg Q60650)

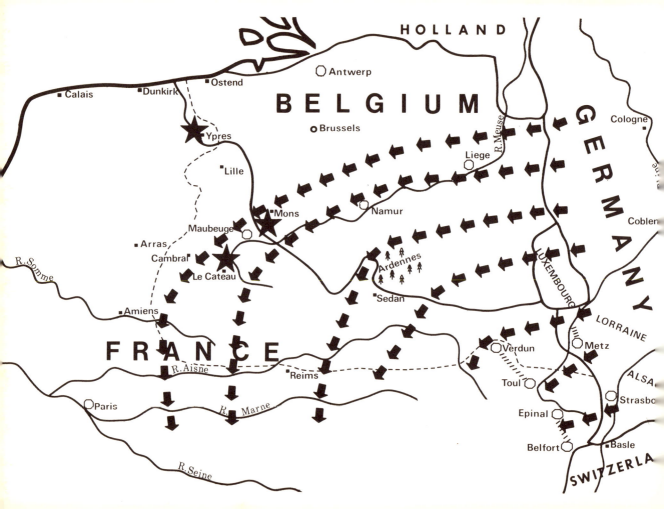

KEY

 ADVANCE OF GERMAN ARMIES

 BRITISH ACTIONS 1914

◯ FORTIFIED AREAS

_ _ _ _ FRONT LINE, WINTER 1914/15

THE SCHLIEFFEN
PLAN AND THE
WESTERN FRONT
BY THE WINTER
OF 1914–15
Although the
Germans failed to
invest Paris and
destroy the French
Army in the opening
campaign, they still
retained the important
industrial areas of
Northern France.

district and concentrated at a given assembly point. The Territorial Army and its Reserve, composed of men between the ages of thirty-seven and forty-seven, were to form units corresponding with those of the field army and the Reserve. The Territorials were also to carry out garrison and depot duties.

Amongst the *élite* troops of the French Army were the African units. These consisted of four regiments of *Zouaves*, four regiments of Algerian riflemen, or *Turcos*, and ten light cavalry regiments; in turn comprising six regiments of *Chasseurs d'Afrique* and four regiments of *Spahis*. The *Turcos* and *Spahis* were coloured troops commanded by French and native Officers.

Infantry tactics and training were based on the principle of the offensive's superiority and on the winning of ground by fire and forward movement only. The role of firepower was relegated to the mere preparation of ground prior to the resumption of forward movement and culminating in the bayonet charge. The same doctrine also regulated the employment of artillery. Small, mobile four-gun batteries of 75-mm quick-firing guns were relied upon to provide sudden, brief and vicious barrages. Gunners were not adequately trained in the use of the '75' beyond a range of 4,000 yards, a factor which tended to negate the technical superiority of the weapon.

21

By 1914, France had failed to equal German military expansion, man for man, while her concentration upon the morale element had led her to ignore the influence of firepower. The reliance of French military doctrine upon semi-Napoleonic tactics in the face of modern, massed rifle and machine-gun fire, and the consequent neglect of equipment and training techniques more suited to twentieth century warfare, undermined the strength of the French Army.

British cavalry during the retreat from Mons in August 1914. (IWM Neg Q60705)

3 August-October 1914: The War Plans Unfold

During the first month of the war, the most prominent feature of military operations was the initial success of German operational plans in seizing and maintaining the strategic and tactical initiative. A situation of imbalance, favouring Germany, was created in the West by the rapid concentration and advance of superior numbers in a wholly unexpected and relatively unguarded area. This was the essential element in the war plan developed by Count Alfred von Schlieffen, Chief of the German General Staff, and amended by his successor, Colonel-General Helmuth von Moltke. A total of 1,500,000 troops had been massed in the West to eliminate France from the war in the first six weeks of operations. Once France had been defeated, it was envisaged that the bulk of these troops could be switched to the East to meet the more slowly developing Russian offensive.

Schlieffen's recipe for success against France consisted of a broad, massed sweep by the German right wing through Belgium and Northern France, by-passing the exposed left flank of the French Army. As the French made their expected offensives against well-defended positions in Lorraine and the Ardennes, the German right wing would encircle Paris from the West and South. The culmination of this vast envelopment would be the squashing of the French armies against their own Eastern frontier.

The boldness of this strategic concept was apparently endorsed by the fall of

Brussels on 20 August, the jarring repulse of French offensive efforts in the Battles of the Frontiers from 14 to 25 August and the general retreat of the Allies along the whole offensive line. However, the dwindling impetus of the preliminary strategic surprise was eroded by the unexpected resistance of the Belgian Army and delays to the timetable of advance by rear-guard actions such as Mons (23 August), Le Cateau (26 August) and Guise (29 August). As the thin veil of strategic originality evaporated, the military realities of the campaign revealed the unimaginative tactics and logistical bankruptcy upon which the German plan was obliged to rely. The deep penetration of the armies of the right wing into northern France emphasized their reliance upon the lines of communication. These highly sensitive, life-giving cords regulated the speed and efficiency with which a mass army could advance. This contrasted with the rapidity with which the enemy was able to transfer troops by rail to meet an unmechanized offensive. Thus as General Joffre, the French Commander-in-

French infantry in September 1914. (IWM Neg Q60729)

Chief, began to appreciate the true nature of the German attack, he was able to reinforce and gather troops on the Allied left. Many of these troops were switched by rail from the Eastern frontier, despite an unscheduled German attack in that area, which was designed to bring about the double envelopment of the French Army.

The early tactical success of the German Army had failed to defeat decisively the Allied armies which still maintained both their will and capability to fight. Furthermore, the deceleration of the German offensive caused a loss of mutual cohesion amongst the advancing armies. This factor was partly responsible for the German decision not to encircle Paris, as had been originally planned. Instead Colonel-General Alexander von Kluck, commanding the German 1st Army, took the risk of wheeling his troops inwards to the east of Paris. In so doing, the right flank of the German Army was exposed, guaranteeing the probable success of an Allied counterstroke. In military terms, the campaign was already a failure for the Germans. They had not defeated and destroyed the enemy in the field. Moreover, they had relinquished the initiative by allowing the enemy to concentrate his forces in an area where he posed a very real threat to the German Army itself. These failures were directly attributable to the Schlieffen Plan's fundamental weakness in ignoring the existence of the enemy's logistical resources, particularly his railways, as a real military target.

As the exhausted German armies dangled by their tenuous lines of communication in front of Paris, the Allies took their opportunity and counter-attacked along the Marne on 6 September. A thirty-mile gap, covered only by a cavalry screen, was opened between the 1st Army, under von Kluck, and the 2nd Army, commanded by Colonel-General Karl von Bülow. The threat of an advance into this gap by the British Expeditionary Force led Bülow to order the retirement of his own army on 9 September. Kluck fell back on the same day and, by 11 September, the retreat had extended to all the German armies. However the reticence of the Allied pursuit enabled the Germans to fall back north of the Aisne, where they entrenched on 12 September. During the next two days, Allied assaults failed to dislodge them; the war of manoeuvre was being replaced by the immobile dominance of defensive firepower.

As an alternative to what was believed to be only a temporary standstill, both sides made a succession of attempts to turn the other's open Western flank. Inappropriately termed 'the race to the sea', the consecutive outflanking movements shifted the main battleground from France to Flanders. The fall of Antwerp on 10 October released more German troops who were to co-operate with four fresh corps in sweeping down the coast and outflanking the Allied left. However, the Belgian Army was successful in holding the Germans on the coast. After ten days of dogged resistance on the Dixmude-Nieuport line, King Albert of the Belgians approved the opening of the sluicegates at Nieuport and the flooding of the countryside. This forced the German flanking movement inland where it collided with a simultaneous

French cavalry in support near Ypres, October 1914. (IWM Neg Q60711)

British attack in the Ypres region. It was here that the war of movement finally spluttered and died. The numerically inferior B.E.F. was thrown onto the defensive from the beginning. However this disadvantage was largely offset by the benefits to be gained from combining a high standard of musketry with even the most primitive defensive position. The British held on grimly to their slender lines in front of Ypres, despite vigorously renewed German attacks on 31 October and 11 November. On the latter date, the Prussian 1st Guard Brigade broke through north of the Menin road, but was driven off by a composite force of cooks, engineers, batmen and the 2nd Oxfordshires. By mid-November the crisis at Ypres had passed but the cost had been heavy. Out of the original 160,000 men of the B.E.F., 86,237 were casualties, over 50,000 of whom had fallen during the Battle of Ypres.

The failure of both sides either to outflank their enemy or to break his front line by direct assault, was an expression of the economic, military and psychological balance which existed between the opposing nations. The result was a static line of trenches drawn from the Belgian coast to Switzerland. Onto the narrow strip of soil which separated the armies, the most powerful nations in Western Europe were to pour their resources of men, materials and ingenuity in an effort to achieve the decisive victory.

Cavalry pass to the rear as the machine-guns are moved to the front during the First Battle of Ypres. (IWM Neg Q60752)

4 The Defensive System

During the harsh winter of 1914–15, both sides concentrated upon establishing a continuous defensive front by linking unconnected positions together with trenches and breastworks. This consolidation, which helped to ease the tactical problems of local defence, was viewed as a purely temporary measure. The trenches were considered to be an uncomfortable but temporary winter billet, where the exhausted armies could gather their strength in anticipation of the return of mobile warfare with the initial offensives in the spring of 1915. In the meantime, both France and, particularly, Germany were able to utilize the equipment they had developed for possible siege operations against fortresses, in the defence of their trenches. On the other hand, the B.E.F.

British and German officers exchange seasonal greetings on Christmas Day 1914. (IWM Neg Q50719)

was almost wholly deficient in heavy guns and howitzers, high explosive shells, trench mortars, sandbags, signal pistols and periscopes. From the beginning, the British were obliged to improvise everything. For example, hand grenades were made of jam tins and mortars of field-gun cartridge cases. Between 1912 and 1914, the British Army had held only one practice exercise in the attack of defensive works and the General Staff had paid no attention to the tactical lessons of the siege and defence of Port Arthur during the Russo-Japanese War (1904–5). On manoeuvres, infantry were discouraged from digging trenches by the rule that they must return afterwards and fill them in.

During the first phase of the fighting

in France and Flanders, the troops had often thrown themselves into hastily scratched-out rifle-pits or any convenient ditch and field drain in order to escape the unprecedented intensity of small arms and artillery fire. During the winter, these isolated ditches and shallow trenches, sometimes separated by as much as 200 to 400 yards, were connected and strengthened by further digging. Where it was possible, the trenches were deepened and the excavated earth thrown on top to form a parapet. As the trench system became permanent, they

Germans digging trenches in the Argonne during 1915. (IWM Neg Q45584)

were widened to about seven feet at the top and two feet at the bottom, with a depth of six feet six inches to seven feet. In winter months, the drainage of the trenches posed a particularly difficult problem. Timber, wattling and sand-bags were used in the construction of revetments to strengthen the sides of the trenches. Sandbags were the most common material used since they required fewer specialist skills, were safer under shell-fire and easier to repair. In the low-lying regions of Flanders, the construction of deep trenches was impossible since the water-table was reached at a depth of about eighteen inches. Here, the trenches consisted of breastworks of sandbags and wood, raised above ground level to a height of six to seven feet and, in some cases, were more than eight feet thick. In the British lines, adequate overhead cover was not

British front line trenches at Laventie in December 1915. (IWM Neg Q17398)

adopted for some time and, as a result, living accommodation tended to be of the most primitive nature. At first, shelter for the Other Ranks usually comprised a hole scraped out of the front parapet or trench wall, sometimes it was simply a waterproof sheet. These 'funk holes' were eventually enlarged and roofed with wooden planks or corrugated iron with earth thrown on top for added protection. It was not until the summer of 1916 that a satisfactory shell-proof cover was generally available. This contrasted sharply with the deep, well appointed, concrete dug-outs constructed for the German troops from 1915 onwards. This simple factor reflected the fundamental difference of tactical attitude which separated the two sides. In fighting on French and Belgian soil, the Allies were on the strategic defensive; a situation which could only be reversed by means of the tactical offensive. The Germans, on the other hand, were happy to maintain their strategic initiative by the simple expedient of a tactical defensive. Thus the British and

A divisional motor convoy. (IWM Neg Q849)

A battery of British 9.2-inch Siege Howitzers captured by German troops. (IWM Neg Q47664)

French tended to regard the trenches and the trench system as a temporary military aberration which would inevitably be broken by the power of their offensive efforts. Therefore they considered there was no necessity to provide sophisticated accommodation, which might dull the aggressive instincts of their troops, in what was only a temporary situation. It was, consequently, left to field units and individuals to provide their own level of comfort and protection, and it was through their almost imperceptible efforts that the permanent trench system was created and solidified.

The trench system adopted by the British, normally consisted of three lines comprising the front, support and reserve trenches. Each of the trenches

A German machine-gun position in the Argonne in 1915. (IWM Neg Q45396)

was constructed as a series of traverses (right-angled zig-zags) which gave protection from flanking fire and minimized the local effect of shells. The three lines, usually separated by a distance of 150 to 450 yards, were linked by traversed communication trenches. In some areas, however, this theoretical standardization was never achieved because of the unabated intensity of the fighting or the geographical and climatic conditions of the region. The problems of establishing an integrated system of trenches were further aggravated by the British High Command's decision not to allow front line units to relinquish ground voluntarily. It was, therefore, difficult for battalion commanders to improve their tactical position or regulate their front line if it involved yielding even a few hundred yards of muddy, shelltorn dirt. For this reason the British line was indented by German salients and strongpoints, usually situated on higher ground such as at Passchendaele and Vimy Ridge. From these positions the Germans were able to command whole stretches of the British front with flanking fire and, at the same time, observe all movement in the rear areas. The only military response made by the British commanders consisted of costly and ineffective attacks designed to take out the salients and strongpoints by frontal assault.

For both sides, the strength of the trench system was its fire-power. The outstanding tactical lesson of the war was the unassailable ascendancy of the heavy machine-gun as the foremost defensive weapon. The water-cooled Maxim variants and the heavy-barrelled, air-cooled French Hotchkiss machine-guns dominated the battlefield from their permanent emplacements. These weapons, when grouped together, could provide a series of interlocking arcs of almost continuous fire. Already indispensable in defence, the machine-gun was used, as the war progressed, to support infantry attacks by laying down an unsighted barrage of fire on predetermined enemy positions. However the size of these weapons and the difficulties of supplying them with enough ammunition during operations, meant that they were too cumbersome to accompany an attacking force. This

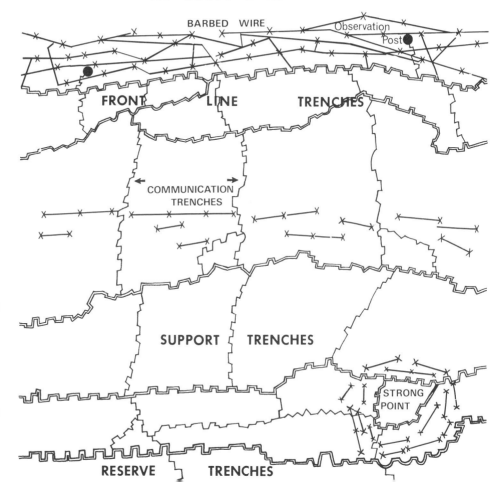

NO MAN'S LAND

BARBED WIRE

Observation Post

FRONT LINE TRENCHES

COMMUNICATION TRENCHES

SUPPORT TRENCHES

STRONG POINT

RESERVE TRENCHES

THE TRENCH SYSTEM
During the winter of 1914–15, both sides concentrated upon establishing a continuous defensive front by linking unconnected positions with trenches and breastworks. This improvised building created a defensive system which could not be broken by conventional infantry assault.

emphasised the need for light, portable machine-guns which could be carried by one man. Probably one of the most versatile of these weapons to be developed was the Lewis gun, designed by an American called Isaac Lewis, in 1911. This light machine-gun, which had a bipod mounting and circular drum magazine fitted above the barrel, was extensively used by British front line troops for close support fighting and was also fitted to a number of aircraft.

The general features of the defensive system were completed by the belts of barbed-wire which were sown by both sides in No-Man's-Land. Since the distance between the opposing lines was as little as twenty-five yards in some areas, although the average was about 250 yards, the wiring of a position became an essential element in blunting the enemy's offensive efforts.

Barbed-wire entanglement and shell holes at Beaumont Hamel in 1916, presenting an almost insuperable barrier to the infantry attack. (IWM Neg Q1547)

German barbed-wire near Arras in June 1917. (IWM Neg Q2548)

5 Artillery

In August 1914, there was an approximate parity in the artillery organization and equipment of the three major combatants. The Germans had an initial advantage in possessing a number of howitzers capable of being used in both field and siege operations. Probably the most widely publicized of these weapons were the 42-cm 'Big Berthas' used to reduce the Liége forts. In contrast, British and French tactical doctrine emphasized the mobility of field pieces to the cost of range and firepower. Generally, both sides failed to estimate accurately the daily consumption of their artillery pieces once engaged in combat. Secondly, none of the combatants had investigated the effect of developments such as railways, aircraft, telegraphs, telephones, radio and survey on artillery technique.

During the first two months of the war, both sides soon came to appreciate the importance of preparing infantry attacks with an artillery bombardment. In defence too, battles such as Le Cateau stressed the need to distribute and conceal artillery deep inside the defensive line. The early experience of trench warfare emphasized the role of howitzers which could lob heavy shells behind breastworks and into trenches. Furthermore, the capacity of high explosive shells to smash defensive positions and cut barbed-wire soon became apparent. Shrapnel shell, which was ideal for use against troops in the open, took on a subsidiary role, even amongst the field-gun batteries where it had been used exclusively. In order to fire heavier and more destructive shells, there was an expansion in the proportion of medium and heavy guns in each army. In British service the 60-pounder and 6-

An 18-pounder field gun embedded in the Flanders mud. The massive artillery bombardments on the Western Front created a muddy barrier across which it was virtually impossible to supply even slow-moving advances. (IWM Neg Q6236)

inch guns and the 6-inch, 8-inch and 9.2-inch howitzers became the modern equivalent of the battering ram.

Early experience during 1915, particularly at Neuve Chapelle, indicated the effectiveness of the preliminary artillery bombardment as the infantry swept past the demoralized defenders and pulverized positions. Thus, as defences became more sophisticated, barrages lasted longer and increased in intensity. However, by 1916, German defences

A shell dump near Contay. (IWM Neg Q3245)

A 6-inch 26-cwt Howitzer in the mud near Pozieres in 1916. Before the war, both sides under-estimated the importance of heavy and medium guns and howitzers. (IWM Neg Q1490)

were so strong that there were always sufficient troops and machine-guns left after the bombardment to repulse an infantry assault. Furthermore, the enormous artillery preparations of 1916 and 1917 continually sacrificed the element of suprise in the mounting of offensive operations. On many occasions, the artillery failed in its primary task of destroying the enemy barbed-wire. At the same time, the thousands of tons of shells created a pockmarked landscape of mud and craters across which it was virtually impossible to supply even slow moving advances.

During the mid-war years, 'refinements' in artillery tactics tended to stress the expanded use of the instrument. Creeping barrages, providing a moving curtain of protection in front of advancing troops, and heavy counter-battery fire became the conventional overture to offensive efforts. On the other hand, by 1917, the German system of defence in depth, with the bulk of the defending force out of artillery range, nullified the effect of offensive barrages. More significant was the employment of artillery in the de-fence to break up impending attacks. This usually took the form of shelling enemy batteries, stationary barrages on No-Man's-Land and the bombardment of enemy front line trenches where troops were gathering for the assault.

It was not until 1918 that artillery contributed directly to the success of an offensive on the Western Front. By that year, the Germans had correctly gauged the effect of the brief hurricane bombardment of only several hours duration against an enemy's defensive system. Using tactics first tried at Riga in September 1917, the Germans pounded the Allied line in March 1918 with high explosive, gas and smoke shell. During a five hour bombardment, fire shifted back and forth against selected targets within the defensive zone. The result was a total disorganization of enemy communications and the smashing of strong points. Later in 1918, the tank displayed that it could adopt several of the functions of artillery. It could demoralize enemy troops, level barbed-wire and eliminate pockets of resistance.

To cope with the expansion in the use

Shells for the German 38-cm railway gun 'Long Max'; this gun had a range of nearly thirty miles. (IWM Neg Q44809)

of artillery, considerable developments were made in artillery technique and organization. Fire control was improved by the use of observation aircraft to range guns onto their targets. Inter-communication between batteries and spotters was improved by the use of field telephones and radio. Surveying, sound-ranging and flash-spotting similarly reflected the growing sophistication of artillery methods. In terms of administration and organization, both sides appreciated the need for a centralized artillery command to co-ordinate action, concentrate fire and economize resources. In addition, the need for a centrally controlled reserve, particularly at Corps level, led to the establishment of units outside the regular formations. Thus, field-guns were rarely distributed above divisional level whilst medium and heavy artillery were the preserve of the Corps.

6   1914-1915: The Strategic Alternatives

By the winter of 1914–15, the deadlock on the Western Front was complemented on the Eastern Front by the failure of both sides to achieve a decisive victory. On the outbreak of war, the Russians had undertaken a direct attack upon German territory by invading East Prussia. This Russian thrust, which developed with unexpected speed, was designed to relieve German pressure on the French armies in the West. Although partially successful in helping the French, the two invading Russian armies were heavily defeated in August and September 1914 at the Battles of Tannenberg and the Masurian Lakes. The effect of these victories was diminished on the Galician Front where an offensive by the Austrian armies into Poland was forced back

in disorder by a Russian counter-attack. As the weather improved in early 1915, the Russians renewed their offensive against the Austrians in the Carpathians. However, a combined German and Austrian attack at Gorlice on 2 May launched on a twenty-eight mile front broke through the Russian line to a depth of 100 miles in two weeks. The Russians were forced to abandon Galicia and most of Poland, losing Warsaw on 4 August. By September, the Russians had been pushed back to a line which ran from Czernowitz on the Rumanian frontier in the south to Riga on the Baltic. By the end of 1915, the Russians were staggering, although they were not yet out of the war.

The war of rapid movement and dramatic defeats in the East, fought

The shell-swept road to Guillemont in September 1916. (IWM Neg Q1163)

across a geographically vast arena, heightened the contrast with the stagnation of the Western Front. However, in late 1914 and early 1915, it appeared to the Allies that a new flank could be created by the invasion of Turkey, which would ultimately threaten the entire position of the Central Powers in Europe. The origins of the Allied attempt to force the Dardanelles Narrows and threaten Constantinople resulted from a Russian appeal, on 2 January 1915, for a diversion in the Eastern Mediterranean which would distract the Turks from their efforts in the Caucasus theatre against the Russians. A direct attack upon Turkey offered the triple opportunity of removing an important German ally, opening the vital sea route to Russia and persuading hesitant Balkan neutrals to join the Allied cause. Kitchener, the Secretary of State for War, was reluctant to commit troops to such a scheme and suggested instead a naval demonstration. Winston Churchill, then First Lord of the Admiralty, advocated the idea that the forts guarding the entrance to the Dardanelles could be reduced by a force of obsolete battleships and cruisers. From 19 February to 15 March, 1915, the Allied fleet intermittently bombarded the forts at the entrance to the Straits. The main attack on the more formidable defences inside the Narrows began on 18 March, but the sinking of three battleships and the damaging of a further three by mines resulted in the naval effort being abandoned. Evidence has since suggested that the Turkish defenders were on the point of collapse and that a determined drive through the Straits would have brought Constantinople under Allied gunfire. Instead the British naval commander, Vice-Admiral de Robeck, advised London that further progress was impossible without military aid.

During February and March, troops had been made available and hastily formed into the Mediterranean Expeditionary Force under the command of General Sir Ian Hamilton. The haphazard preparations of the army at Mudros on the island of Lemnos, culminated in a landing on the Gallipoli peninsula at Cape Helles on 25 April. Since February, the Turks had built up their available forces on the Straits from two to six divisions and were successful in containing the main landing at Cape Helles and also the two diversionary attacks by the French and the Australian and New Zealand Army Corps (Anzac). A lack of firm centralized control of the Allied forces left them stranded on congested beaches with the Turks everywhere holding the heights overlooking the landings. The subsequent attempts to get clear of the beaches ended in failure and the conflict dissolved into conditions of trench warfare. A new landing at Suvla Bay on 6 August was unopposed by the Turks but instead of pressing inland the troops were ordered to dig-in on the beach. The Turks were unable to make any appreciable addition to their strength in the area for over forty-eight hours after the original landing. However, due to inexcusable British delays, they were able to win the crucial race for the heights which dominated the British position. Probably the only real Allied success of the Gallipoli Campaign was the evacuation of the whole peninsula which was carried out during December and January 1916 without the loss of a single man from enemy action.

The failure of the Gallipoli expedition appeared to vindicate those military leaders who argued that decisive victory could only be found on the Western Front. The idea that the war could be won on the cheap somewhere other than in France and Belgium was certainly discredited for the remainder of the war. The unfortunate irony was that the factors which had mitigated against success at Gallipoli were also obvious on the Western Front.

7   Offensive Tactics

British Infantry Tactics

Throughout the war, the onus was upon the Allies to recover the ground which had been lost in 1914. Thus, from 1915, the British and French adopted a programme of unrelenting offensive effort which was aimed at eroding the enemy's defensive strength. The practical reality of this policy of attrition consisted of a series of frontal assaults made by infantry supported by artillery. The tactical pattern for future British offensives was established at Neuve Chapelle in March 1915. Following thorough preparations by staff officers and an intense thirty-five minute bombardment, the assaulting battalions quickly overran the shattered front trenches of the German centre. However difficulties on the flanks and congestion in the rear areas deprived the attack of its impetus.

Also, delays in collecting troops for an attack on the enemy's second line gave the Germans an opportunity to redress the numerical disparity by bringing up reserves. When the battle ended, the British had advanced 1,000 yards on a 4,000 yard front for 13,000 casualties in five days. Yet a substantial area of the ground gained, including the village of Neuve Chapelle, had been won during the first three hours of the battle.

British commanders attributed the deceleration and collapse of the assault at Neuve Chapelle to a lack of men, munitions and equipment. On this basis, the belief took root that only heavier artillery preparations and a massed assault could sustain an offensive effort that would level the enemy's defences and break his reserves. This in-

British troops occupying captured trenches on the Somme in 1916. The casualties incurred in taking a few yards of enemy trench usually made it impossible to continue the attack. (IWM Neg Q3990)

creasing emphasis upon the importance of sheer physical density in attack, disguised the tactical ineptitude which was the usual hall-mark of the massed offensive. The casualty lists of successive offensives, such as Loos (1915), The Somme (1916) and Passchendaele (1917), revealed the insolvency of British infantry tactics.

The emergence of the theory of mass corresponded to a decline in both field-craft and the encouragement of tactical initiative amongst attacking units. Furthermore the deployment of large numbers of raw volunteer troops from late 1915, also discouraged the evolution of imaginative and realistic infantry tactics. In the attack, the battalion advanced across No-Man's-Land in a series of successive waves, each wave consisting of troops extended at about one man every two yards, and with about fifty yards between waves. Within the battalion, each of the four companies advanced in two waves. The first wave comprised two platoons abreast covering a front of 200 yards. Each platoon consisted of two lines separated by fifteen to twenty-five yards.

The first line was made up of riflemen and bombers and the second of Lewis gunners and rifle bombers. They were followed by 'clearing' parties, composed largely of bombers whose task was to mop up enemy resistance in the captured trenches. However, advancing in serried ranks, the waves presented easy targets for German machine-guns and they rarely reached even the preliminary objectives.

The basic unit of the assault was the platoon which was usually assigned a limited tactical objective within the enemy trench system, such as a strong point or the junction of a communication trench. The *Instructions for The Training Of Platoons for Offensive Action, 1917*, in summarizing the tactics to be used in the attack, emphasized that platoons push on to their target at all costs and 'get in with the bayonet'. This blunt directive denied the platoon the flexibility and fluidity which are the essence of infantry tactics. Even if the objective was secured, instructions demanded that the position be consolidated. Similarly an over-sensitivity about the security of flanks meant that an ad-

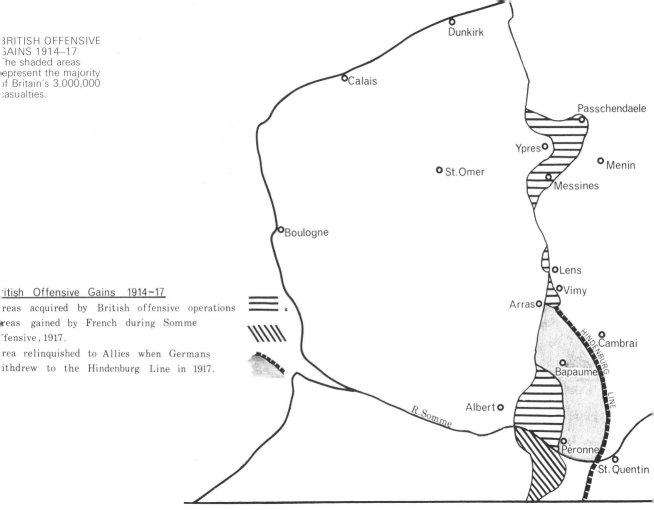

BRITISH OFFENSIVE
GAINS 1914–17
The shaded areas
represent the majority
of Britain's 3,000,000
casualties.

British Offensive Gains 1914–17

areas acquired by British offensive operations
areas gained by French during Somme
offensive, 1917.

area relinquished to Allies when Germans
withdrew to the Hindenburg Line in 1917.

Dunkirk

Calais

Passchendaele

St.Omer

Ypres

Menin

Messines

Boulogne

Lens

Vimy

Arras

HINDENBURG LINE

Cambrai

Bapaume

Albert

R.Somme

Péronne

St.Quentin

vance at brigade strength could not be made until all the opposition had been cleared. Thus the existence of several uneliminated pockets of resistance could stall even a large assault.

In theory, each offensive consisted of three tactical phases. The first was artillery bombardment, followed by the infantry assault and breakthrough. Finally the cavalry were to exploit the gap in the enemy's line created by the infantry and re-instate the war of move-ment. However on the Western Front, cavalry were not to be employed in their allotted offensive task. This was largely due to the fact that the infantry attack rarely advanced beyond the enemy's first line of defence. Until the introduction of the tank, commanders correctly assumed that cavalry offered the only means of rapid mobility. However their vulnerability and inability to cope with trench warfare conditions rendered them tactically inappropriate.

Chemical Warfare

The scientific and industrial developments of the major combatants made the adoption of chemical agents as military weapons an inevitable step. The most common form of chemical warfare was the use of gas to paralyse the enemy's forces or make his lines untenable. The first incidence of its use was at Neuve Chapelle in October 1914, when the Germans fired shrapnel shells treated with an irritant substance. In the following January, a number of shells containing xylyl bromide (tear gas) were

Sentry ringing a gas alarm. (IWM Neg Q669)

employed against the Russians at Bolimow on the Eastern Front. The first significant use was at dusk on 22 April 1915, when chlorine gas was released against British and French positions in the Ypres Salient. A five mile gap was created in the defensive line and it was only through the failure of the Germans to exploit their initial success, that a major Allied set-back was avoided. A lack of German reserves and tactical unfamiliarity in the use of the weapon gave British and French forces the opportunity to recover from their surprise and restore their front. Within a few days, Allied troops had been issued with pads of cotton waste which were to be dipped in bicarbonate-of-soda solution and tied over the nose and mouth in the event of a gas attack. These crude masks were continually improved during the war. By 1918, both sides were equipped with respirators incorporating charcoal filters and chemical solutions to neutralize the gases. Thus the premature use of gas in April 1915, robbed it of its decisive potential and for the rest of the war it became a mere tactical accessory.

At Ypres, the Germans had used

German troops practising with a portable flamethrower. (IWM Neg Q55559)

cylinders to release the gas, relying upon the wind to carry it into the enemy trenches. This form of delivery limited the number of chemical gases which could be used and was also rather unreliable. During the first British gas attack, at Loos on 25 September 1915, a shift in the wind blew the gas back into the British trenches. Therefore, from 1916, both sides made increasing use of gas-filled artillery, mortar and projector shells. This development also widened the choice of toxic chemicals which could be employed. In mid-1917, the Germans began filling shells with an almost odourless agent known as mustard gas. As it caused severe blistering to the skin and respiratory tract several hours after exposure, adequate protection was difficult. Furthermore its high persistency rate in heavy soils could render areas untenable for several weeks. Mustard gas caused the majority of gas casualties followed by chlorine, phosgene and chloropicrin. The latter three depended for their effect upon damage to respiratory organs. The French made limited use of prussic-acid gas, which usually attacked the victim's nervous system.

By March 1918, a concentrated bombardment of enemy positions with gas shell was a vital preliminary to a German infantry attack. The normal method was to direct mustard gas onto the enemy's flanks and a less lethal, low persistency gas against the front selected for attack. To meet this demand, the German shell production programme for 1919 estimated that 50% of the ammunition manufactured would contain gas. Similarly, the British intended to fill about 33% of all artillery shells with toxic chemicals.

In addition to gas shells, both sides made extensive use of smoke, incendiary and star shells for local tactical effect. Another chemical weapon, introduced to the battlefield by the Germans in 1915, was the flamethrower. This had first appeared in the early 1900s when the German Army tested two models submitted by Richard Fiedler. By 1911, three pioneer battalions had been issued with *flammenwerfer*. The two patterns adopted consisted of a small portable version and a larger static type. Both used gas pressure, usually nitrogen, air or carbon dioxide, to force a stream of oil through a small nozzle where it was ignited. The incendiary range of the light pattern was about twenty-five yards whilst the heavier model could fire for nearly forty seconds over a distance of forty yards. None of the flamethrowers developed and used during the war had sufficient range or duration of fire to be tactically effective. Their influence seems to have been limited to terrorizing the troops they were used against.

Tunnelling and mining

The defensive deadlock of the Western Front reproduced, on an unprecedented scale, the characteristics of siege warfare. In the British Army, no attention had been paid to the examples of attack and defence of field fortifications by mining during the Russo-Japanese War. It was, therefore, the Germans who took the initiative in late 1914. They began tunnelling operations which culminated in the blowing of several mines under the British lines at Givency, Guinchy and in the Ypres Salient. From mid-February 1915,

specially organized British tunnelling companies began moving into the line. It was indicative of British unpreparedness that the only smoke helmets available for these men were relics of the Crimean War. By the middle of 1916, there was a total British force of 25,000 men engaged in mining operations. This was largely due to the efforts of Major J. Norton Griffiths, MP, who had been authorized to enlist miners for service in France. During 1915, British operations were restricted to holding the German tunnelling offensive. However, by 1916, an improved central administration, better equipment and the co-ordination of mining activities contributed to the opening of a permanent British offensive. As mine warfare reached its peak in 1916, over thirty miles of the eighty mile front occupied by the British were protected by underground galleries. During that year, 1500 mines were fired along the British-German front.

The normal tunnelling system consisted of two parallel shafts, or inclines, driven to the required depth and from which the main galleries were driven. The main galleries were connected by a

lateral tunnel which gave ventilation
and security. Fighting galleries and
listening-posts were built out from the
lateral. Other tunnels were driven out to
protect the flanks of the main gallery
and the lateral. The success of mining
operations depended upon getting be-
neath the enemy's tunnels and rapidly
destroying his counter-mines. This was
usually achieved by the use of a cam-

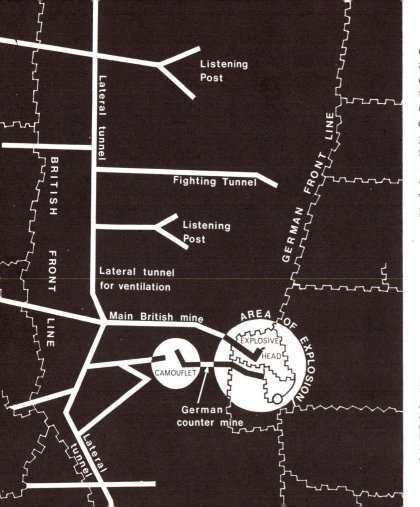

Listening Post

Lateral tunnel

Fighting Tunnel

Listening Post

BRITISH FRONT LINE

Lateral tunnel for ventilation

Main British mine

AREA OF EXPLOSION

GERMAN FRONT LINE

EXPLOSIVE HEAD

CAMOUFLET

German counter mine

Lateral tunnel

ouflet or a 'torpedo'. The former was a small mine designed to destroy the enemy's workings substantially. The torpedo consisted of an explosive charge pushed along a small tunnel close enough to collapse the German gallery when it was exploded. The culmination of British operations came on 11 June 1917, during the Messines ridge offensive. Altogether nineteen mines were fired containing nearly a million pounds of high explosives.

MINING AND
TUNNELLING
Although mining was
an important feature
of siege operations,
it was an expensive
and laborious way of
gaining ground.

A sectioned Mills grenade; when the pin was pulled and the grenade thrown the pressure on the lever was released and it flew off. This allowed the striker to be forced down by the spring onto the percussion cap which exploded. This, in turn, lit the fuse which burnt for five seconds and then ignited the detonator. The detonator then exploded the high explosive contained in the charge. (IWM Neg Q58163)

8 Weapons and Equipment

Grenades and Mortars.

The proximity of the opposing trenches and the protection which they provided against small arms fire prompted the introduction of specialized weapons. Probably the most important development was the vast expansion in the employment of hand grenades and mortars. In August 1914, the only British soldiers to be trained in the use of grenades were the Royal Engineers. However, from May 1915, a serious programme of training infantry as bombers was undertaken. When the Germans had made their first bombing attack against the British in September 1914, the B.E.F. only had one type of grenade available. Because of the manufacturing costs, however, it was rarely obtainable at the front. This led to the improvisation of grenades by infantry

and engineer units in the field. Jam or milk tins were filled with ammonal or guncotton and fitted with a short fuse which had to be lighted before throwing. This primitive device marked the introduction of time-fused grenades which were eventually to supercede the percussion type. The most important service grenade was the Mills Bomb; over 33,000,000 were issued from the spring of 1915. A time-fused grenade, its oval body was cast-iron weakened by longitudinal and transverse grooves which, on detonation, broke up into forty-eight pieces. By the end of the war, the hand grenade had become a vital element in infantry tactics.

At the outbreak of war the German Army had nearly one hundred and fifty trench mortars *(minenwerfers)*. These

German dugouts constructed in a mine crater. (IWM Neg Q41761)

had been specially designed to bombard the opposite slopes of the hills around their defensive positions in the fortified Moselle region. By the end of 1914, these mobile mortars with their high, curved trajectories were accurately shelling Allied positions and supporting infantry attacks from the German front line. These weapons, principally the 7.6-cm light, 17-cm medium and 25-cm heavy *minenwerfers* were precision pieces incorporating rifling and recoil mechanisms. In contrast, the first Allied mortars were more closely related to lengths of domestic pipe. Firing nails and other metal scrap, these improvisations were, on the whole, more dangerous to the user. Even later Allied mortars tended to be cumbersome smooth-bores without any recoil mechanism. The most important development was the British 3-inch Stokes mortar which was the precursor of the modern infantry pattern. To operate the Stokes mortar, a safety-pin was withdrawn from the shell which was then dropped down the muzzle and fired almost at once. By December 1917, there were 3,500 mortars in front line

use with the British Army.

Uniforms and Miscellaneous Equipment
Field conditions demanded that uniform and equipment adapt to meet the requirements of trench warfare. The need for economy of design, practicality and concealment brought alterations in uniform patterns. Until 1915, French infantry wore the red-topped blue-banded cap, blue double-breasted overcoat and red trousers which had been worn during the Franco-Prussian War. Although the uniform reflected the French theory of *élan* in the offensive, it made the infantryman a conspicuous target. In 1915, it was replaced by a uniform of similar design but in horizon-blue. In the German Army, the characteristic spike of the *pickelhaube* was removed by official order in September 1915. At the same time a new uniform pattern was introduced; the simple cut, particularly of the tunic, reflected the shortage of cloth in Germany as well as the need for inconspicuousness in the trenches. British Army uniform, on the other hand, hardly changed throughout the war.

German bombing party training in the use of 'Disc' grenades. The grenades are being removed from a special carrier attached to the dog. (IWM Neg Q55558)

From late 1915, shrapnel helmets were issued by both sides. This was largely a response to the high incidence of head wounds amongst men in the trenches. The helmet was the only standard item of personal protection issued to troops during the war. Suits of body armour were sometimes made available for specialized tasks but remained experimental, never being widely distributed. The problem of producing a cheap, light body armour, effective against small-arms fire, has yet to be solved. Armoured facial protection for

Mortar bombs in a reserve trench dump. (IWM Neg Q4910)

snipers, in the form of steel plates, was quite common on both sides.

In spite of the availability of modern, sophisticated weapons and equipment, trench fighting was distinguished by improvisation and irregular weapons. The most remarkable of these were the clubs, daggers and knuckledusters which were adapted for hand-to-hand fighting in the restricted trench alleyways. These items were usually selected for their manoeuverability, silence and effectiveness. For trench raiding, clubs were essential if prisoners were to be taken alive. The return to medieval warfare was also marked by the use of crossbows and catapults to launch grenades and bombs. However the use of all this equipment was indicative of the unpreparedness of both sides for trench warfare conditions.

9   1916: Strategic Anticlimax

Following the extensive German victories in the East and the failure of Allied offensives on the Western Front during 1915, the German High Command attempted to cripple the French war effort with an all-out offensive. The area selected for the attack was the Verdun Salient. The battle began in February and continued for five months for a total cost of 700,000 French and German casualties. Although Verdun had not fallen to the Germans, the French Army's spirit had been badly mauled. Partly to relieve the pressure on the French at Verdun, the British hurled the volunteer Kitchener armies against the German line on the Somme. The offensive was launched on 1 July and, by the end of the day, the British Army had suffered 20,000 men dead and a

further 40,000 either wounded or missing. When the attacks finally died in the November mud, the British and Germans had each sustained nearly half a million casualties. For all this blood and effort, the British had acquired little except a graveyard for the volunteer armies.

While the armies slogged away on the Western Front, the British and German navies, after two years of shadow boxing, finally collided in the North Sea. Before the war, naval rivalry between Britain and Germany had contributed towards a steady deterioration in diplomatic relations. By August 1914, Britain's capital ships comprised twenty Dreadnought-type battleships and nine Battlecruisers. Germany had fifteen Dreadnoughts and six Battlecruisers.

Mortar shell dump of 'Toffee Apples' in 1916. (IWM Neg Q1375)

71

German body armour.
(IWM Neg Q23945)

British soldiers talking with a wounded German prisoner. The man on the right is carrying a rifle fitted with wire cutters. (IWM Neg Q809)

At the beginning of the war, Germany made a number of hit-and-run raids against eastern English coastal towns and the British fishing fleet on the Dogger Bank. However she was reluctant to risk her numerical inferiority in a full-scale battle with the British Grand Fleet. On the other hand, Britain was unwilling to endanger her margin of superiority by a vigorous offensive, preferring instead to suffocate Germany's war capacity by a naval blockade. In 1916, the British attempted to surprise the marauding German High Seas Fleet in the North Sea. In the subsequent Battle of Jutland, the British received a severely bloodied nose. The Germans were so shaken that they were only to emerge from harbour on three more occasions during the war. Germany's failure to gain surface control of the sea meant that she had to rely more heavily on submarines to strangle Britain's economic life-lines to the rest of the world.

A French periscopic rifle. (IWM Neg Q69982)

A camouflaged, dummy tank and 15-inch Siege Howitzer. (IWM Neg Q17700)

10 Holding the Line

Living Conditions

As the Allied and German offensives failed to release the paralysing grip of trench warfare, the military organization and administration of both sides was gradually moulded to the routine demands imposed by the garrisoning of the trenches. By late 1915, a division holding a length of front would usually comprise of two brigades in line abreast of each other. The third brigade would be resting several miles behind the front lines. Within the brigade, two of the four battalions occupied the front and support lines whilst the other two went into billets as local reserves. After several days, the reserve battalions relieved the front line units who then returned to billets. During a two week tour of duty, a company would spend about eight days in the front and sup-

port trenches and six to eight days in billets as reserves. After a fortnight the whole brigade was relieved and moved into rest camp for about six days rest.

The relief of units at the front was invariably conducted under cover of darkness. Although it was generally advisable to begin reliefs as early as possible, the intervals between reliefs and the times at which they were carried out was varied from time to time to deceive the enemy. The relieving troops were issued with rations for forty-eight hours and 150 rounds of small arms ammunition. The rations might consist of two tins of bully beef, eight biscuits, a piece of cheese, tea, sugar and a tin of pozzy or jam to be shared between two men. Before leaving their billets, the men were given hot tea and an issue of 'VC mixture' or rum, which was always consumed in the presence of an officer. In some units, the issue of the rum ration was forbidden by the commanding officers on religious and moral grounds. In addition to their own packs and equipment, troops going into the front line were laden with trench stores and materials for the maintenance of the trenches. Each man generally carried a quota of forty sandbags as well as picks and shovels. Other items included duckboards, corrugated iron sheets, barbed wire, disinfectant and bags of charcoal for bogie fires. The relieving units were led into the trenches by guides from the unit to be relieved. On dark nights it was sometimes necessary for each man of the party to hold the bayonet scabbard of the man in front to prevent men getting lost. As the troops moved up, they were still subject to casualties from random small arms and artillery shots. Both sides unleashed sporadic, unaimed machine-gun and artillery barrages against known enemy crossroads and junctions to try and cause casualties at night.

In the trenches themselves a daily

routine was soon established. An hour before dawn the company was called to 'stand to' when each man stood on the firing step with his rifle, bayonet fixed, in case of an early morning attack by the enemy. This ceremony was sometimes the occasion for 'morning hate' when

Members of the 4th Company of the German 27th Infantry Regiment wearing sabots filled with straw as a precaution against trench-feet. (IWM Neg Q51079)

both sides fired off quantities of small arms ammunition at each other until the need for breakfast brought a natural halt. Breakfast was followed by a host of mundane military chores such as filling sandbags, digging latrines and, in winter, draining the trenches. For the most part, however, the day was passed in idleness or catching up on sleep. For sentry duty, the troops were divided into groups of six men under a Non-Commissioned Officer. During the night, two men of the group stood sentry, whilst in daylight only one

German trench *Rupprecht-Weg* in the Lieven area during the winter of 1914–15; the barbed-wire barricade above the trench would be pulled down in the event of an enemy raid. (IWM Neg Q51114)

sentry was provided. One officer per company was always on duty and it was his job to move up and down the company front continually to check that all was correct. One hour before dark there was another general 'stand to'.

At night, the real tasks involved in maintaining and supplying the trenches were carried out. This work included the repair of damaged sections of trench and the erection of barbed-wire entanglements. Scouts and patrols were also sent out to test the strength of the enemy's defences and, in particular, his wire. Where the opposing lines were more than 100 or 150 yards apart, a listening post was established by each platoon in No-Man's-Land. These posts

Draining a reserve trench during the winter of 1914–15. (IWM Neg Q49221)

usually consisted of three men and one NCO, who were posted at dusk and relieved every four hours. Their job was to observe the enemy and, in particular, detect any aggressive moves. The officer on duty was expected to visit the listening posts twice each night.

Also at night, water and ration parties left the front for the rear area to secure the necessary fresh water and provisions from the Company Quarter-Master Sergeant. Before regular water supplies to the trenches were provided, the men had to rely upon water collected in shell holes or cavities in the trench. One improvised method was to dig a deep hole in the floor of the trench close to the front parapet. The hole was then lined with sandbags which acted as a filter. The water which trickled into these holes could then be boiled and drunk as tea or cocoa. The conventional water supply was usually carried into the lines, by hand, in rum jars and empty petrol cans. The most satisfactory method of carrying edible rations was in sandbags.

During the routine course of trench duties, a British battalion usually lost about thirty men each month through death, wounds and sickness. A number of diseases previously unencountered had become synonymous with trench conditions. For example, 'trench feet' were caused by incessant exposure to cold, wet and mud, being particularly prevalent during winter months. In extreme cases, it was possible for the sufferer's toes to become gangrenous. Another manifestation was 'trench

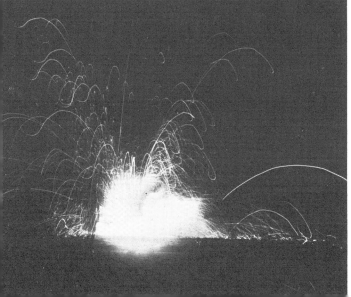

German star shell fired to illuminate enemy positions. (IWM Neg Q445)

79

fever', an infectious disease transmitted by the lice which infested every front line soldier on the Western Front. These tiny insects flourished in the seams of dirty clothing where their eggs were incubated by body heat. However, the majority of casualties were caused by the intermittent shelling, mortaring and sniping which was a daily feature of trench life. Since the majority of troops preferred to spend their tour of duty in relative quiet, they particularly dreaded the arrival of 'travelling circuses' in their line. These were mobile machine-

British soldiers resting in their trenches at Thiepval Wood in August 1916. (IWM Neg Q872)

gun or mortar units, operating from no fixed point, whose task was to annoy the enemy. Once they had moved on, the enemy inevitably retaliated in kind against the defending troops. Similarly unpopular, but less common, were trench inspections by Staff Officers.

The day-to-day defence and maintenance of the front line reflected the way in which the original haphazard cluster of rifle pits had been cemented into a defensive system which imposed its own demands. By 1916, transport, supply, engineering, communication and medical services were no longer geared to the requirements of mobile warfare. Instead of administrative and logistical flexibility, static warfare had spawned a pattern of organizational rigidity which virtually guaranteed the permanency of the trench system.

Women aircraft fitters at work; due to the shortage of labour women took over numerous jobs in both industry and agriculture. (IWM Neg Q27255)

Manpower

In August 1914, Britain had not expected to become involved in a protracted war of attrition fought in Europe alongside and against mass conscripted armies. In terms of reserves of munitions and manpower, Britain had little to offer once the Expeditionary Force had been mobilized and equipped. Apart from the Special Reserve, which was included on the Regular Army's strength of 450,000, the only non-professional reserve units were con-

tained in the Territorial Force of fourteen infantry divisions and fourteen cavalry brigades.

When Field Marshal Earl Kitchener of Khartoum became Secretary of State for War in August 1914, he accurately predicted a long and costly war. Ignoring the Territorials as a nucleus for growth, he advocated a voluntary force built around the Regular Army. The first New Army, of six infantry divisions came into official existence on 21 August. The response to Kitchener's call was overwhelming and, in less than a year, 2,000,000 men had volunteered for service. The support and enthusiasm evinced by the volunteers reflected the emotional nationalism which swept the country during the early months of the war.

Volunteers continued to come forward, at a dwindling rate, after the frenzied days of 1914–15. However, the mounting toll of casualties demanded that 'shirkers' should be sought out and obliged to join the army. In January 1916, compulsory military service was introduced for single men. This piece of legislation represented a break with British military tradition and practice. That it could be passed was an indication of the national commitment to maintaining and eventually winning the war.

The total enlistment of men from all sources in Great Britain came to nearly five million. This represented nearly 25% of the total male population of the country; virtually every household had a member of the family at the Front. Probably no other single factor had ever mobilized the country's social forces in what was believed to be the national interest. One difficulty, however, was the effect which the reduction in available skilled labour had upon industrial production. By mid-1915, it was apparent that it was vital to balance the need for military manpower with the requirement for industrial manpower. In an industrial war, it was more important to maintain the national economic strength and output than to lavish cannon-fodder on the front.

Chateau Wood, near Ypres, in October 1917. (IWM Neg E1220)

Transport and Supply Services

The expansion of the British Army on the Western Front from a ration strength of 120,000 in August 1914, to nearly 3,000,000 by November 1918, posed an unprecedented logistical problem. However the supply services held a fixed stock of advantages which allowed them to keep pace with the army's expansion. The proximity of the United Kingdom as the source of supply served to guarantee an uninterrupted flow of resources. Secondly, the *matériel* was

British Army bread store at Calais in 1917. (IWM Neg Q4796)

Stacks of rations for the British Army at Rouen. (IWM Neg Q1766)

poured into France through large modern towns and ports capable of satisfying the handling and storage demands imposed by such a massive and complex operation. Within France and Belgium, the lines of communication were usually short and always static. Furthermore they were serviced by an existing railway network and metalled roads. The only real impediments to transport were in the forward areas but, even here, the short distances involved usually meant that supplies could be got through.

The Base Supply Depots, such as Boulogne and Rouen, were the clearing houses for supplies and comprised of warehouses, bakeries, cold storage installations and even goat farms. From the depot, supplies were despatched in bulk, by train, to a railway regulating station (railhead) where they were repacked into the solid-tyred, 3-ton lor-

Frozen watering-point during the winter of 1916–17. (IWM Neg Q4853)

Ration wagons on the Somme in 1916. (IWM Neg Q4600)

Frying bacon on a brazier in a reserve trench during March 1917. (IWM Neg Q4840)

horse-drawn wagon to the regimental billets or rest camps from where they were taken, by hand, into the trenches. Probably the most important links in the system were the railways. In 1916, the construction of new rail reached thirty-two miles per day. During 1917, 200 miles of branch line track in the South of England was lifted and relaid in France. By February 1917, 521 British locomotives were operating across 1,000 miles of British-administered track.

The basic cost of food rations for front line troops was calculated at 1/10d (approximately 9½p) per man per day. However there were considerable variations in the types of rations supplied because of the multi-racial nature of the forces involved. Indian troops were provided with *atta* (meal), *ghi* (clarified buffalo butter) and *dal* (pulse) whilst personnel of the Chinese Labour Corps included nut oil in their diet. In addition to the needs of the men, 10,000,000 lbs of forage was required each day for nearly half a million animals.

ries of the Divisional Supply Column. These vehicles carried the supplies to refilling-points. The next stage was by

"YOUR COUNTRY NEEDS YOU"

Regimental cook.
(IWM Neg Q1581)

Kitchener calls for
volunteers. (IWM
Neg Q48378)

A Liven's Projector emplacement; the gas shells are being loaded and electric leads attached to the firing mechanisms. (IWM Neg Q14945)

Communications

Before the outbreak of war, the communication services of the British Army relied largely upon the telegraph and visual signals. By the end of 1914, these resources had proved largely inadequate for the demands of modern warfare. In particular, a new and widening emphasis was being placed upon the need for reliability and efficiency in communications. This was mostly due to the development of artillery and air reconnaissance as well as the growing complexity of military organization and administration in forward areas. The response was to supply the conventional equipment in larger quantities. The line telegraph and telephone remained the bases of the Signal Service with their inevitable accompaniment of miles of cables. These cables were originally laid along the bottom of the trenches or fixed to the trench walls with staples. For protection from enemy artillery and casual soldiers, the cables were buried up to a depth of eighteen to thirty inches. This was later standardized at three foot. In spite of these precautions and the sheathing of the cable in brass,

A British listening post (IWM Neg Q27039)

90

Despatch dog with container for carrying messages. (IWM Neg Q23697)

lead and steel, the cables were always vulnerable to enemy artillery. Furthermore buried systems were local and static, requiring only a minor move to render them inoperative.

The alternative methods of communication were even less satisfactory than cable systems. Visual signalling by flags, lamps and heliographs was wholly unsuited to the conditions of trench warfare. The Army Pigeon Service was extensively used to provide routine communications in the British Army. However pigeons suffered from the disadvantage that they could only operate in one direction. The wireless was quite a new innovation and was to remain neglected in British service during the war.

Although the British communication services could cope with the routine requirements of the trench system; they tended to fall down under the stress of battle. A minimum of mobility demanded a flexible system based upon portable, uncomplicated equipment. Yet even in 1918, an attacking force was still followed by men and wagons laying cable routes from the rear areas to the forward units. All too often, these conventional systems broke down and units had to rely upon the runner, usually at a high cost in casualties.

Messenger pigeons and camouflaged loft. (IWM Neg Q27096)

Officers observing the fall of artillery shells and the information being telephoned back to the battery. (IWM Neg Q5095)

Tending wounded in
a trench on the first
day of the Somme
offensive in 1916.
(IWM Neg Q739)

Medical Services

The administrative rigidity of trench
warfare was reflected in the collection
and evacuation of casualties. The wounded were first brought to the Regimental Aid Post, where the proximity to the
front line meant that only the most
basic treatment could be given. From
here they were moved to Dressing
Stations operated by Divisional Field
Ambulances. No facilities were available at the Dressing Stations for surgery
but wounds could be dressed before
further movement. Selected cases were
evacuated, sometimes by motor transport, to the Casualty Clearing Station
(CCS). The CCS was usually situated
several miles behind the lines and,
ideally, out of enemy artillery range.
These large, semi-permanent stations
normally provided the first facilities for
major surgery. Nurses and orderlies
were also available to examine and
classify wounds prior to treatment. For
the seriously wounded, the CCS was
followed, as soon as possible, by evacuation to one of the large Base Hospitals,
such as at Boulogne, in an ambulance

Digging out wounded
from a Regimental
Aid Post hit by shell-
fire near Zillebeke on
20 September 1917.
(IWM Neg Q5979)

An Advanced
Dressing Station at
Tilloy in April 1917.
(IWM Neg Q2023)

Wounded waiting to
be transported to a
Casualty Clearing
Station during the
Third Battle of Ypres.
(IWM Neg E711)

train. In quiet periods, this system could cope satisfactorily with the flow of casualties. However during major battles, the forward posts and stations became choked with wounded and the Casualty Clearing Stations overloaded. On such occasions, wounded lay in the open for prolonged periods and surgeons had to select for treatment only those casualties requiring a minimum of surgery.

In addition to the perennial problem of rapidly evacuating wounded from the battle area, there were new clinical problems to face. For example, there was a high incidence of infection amongst wounds, caused by high explosive shells. This was due to the shell blast introducing heavily contaminated dirt and soil into what were often already complicated multiple wounds.

The heavy soils of France and Belgium also harboured gas-forming bacilli which caused a lethal infection of wounds known as gas-gangrene. Another relatively new problem was the effect of prolonged exposure in the front line upon a man's mental state. It was only towards the end of the war that this aspect was treated as a medical problem. On the other hand medicines were sufficiently advanced to be able to cope with the conventional ills of a campaigning army, namely typhoid and enteric fever. However there was a stubborn belief in the benefits of antiseptics and amputation, which tended to ignore more recent developments such as blood transfusions and blood coagulants.

Motor ambulances of the 16th Irish Division at an Advanced Dressing Station on the Montauban-Guillemont road in 1916. (IWM Neg Q4246)

Wounded at a CCS awaiting evacuation to a Base Hospital. (IWM Neg Q1217)

11   Intelligence and Observation

Aerial Warfare, 1914-18

By the winter of 1914–15, the cavalry were no longer able to carry out their traditional role of reconnaissance. The result was that the armies became increasingly dependent upon aircraft to supply information on enemy movements. Furthermore the growing use of artillery meant that spotting of enemy batteries and ranging of guns onto enemy targets became important new tasks for the air arms of both sides. It was soon apparent that the destruction of enemy observation machines and the protection of one's own reconnaissance aircraft was a vital element in maintaining the superiority of friendly forces on the ground. This requirement to establish an aerial ascendancy over the opposing air service called for the specialization of aircraft types and roles at an early date. For example, Scouts (special fighter aircraft) were developed which had to be equipped with efficient armament. This implied a forward-firing machine-gun which could be synchronized to shoot through the airscrew arc without demolishing the wooden propeller. On the other hand, air superiority was not simply a matter of technical superiority. Both sides were to show that superior tactics and organization often compensated for inferior equipment.

The first tentative steps towards sophisticated aerial combat techniques were made in the Spring of 1915 when a number of French Morane monoplanes had steel plates fixed to the propellers, allowing a machine-gun to fire through the airscrew arc. In June

A crashed German Albatros DIII. (IWM Neg Q42247)

A Flight of S.E.5As
of No 1 Squadron
R.A.F. over St Omer
in June 1918. (IWM
Neg Q12053)

and July, German *Eindeckers* (monoplanes) equipped with Fokker inter-
rupter gear appeared on the Western
Front. After some initial set-backs, the
Fokkers won overall domination in the
air during the last months of 1915. This
success was largely due to Oswald
Boelcke and Max Immelmann who,
whilst operating together, had laid

S.E.5A Scouts of No 85 Squadron, R.A.F., at St Omer, 21 June 1918. (IWM Neg Q12051)

down the rudiments of air tactics by utilizing the advantages to be gained from height, speed and concealment.

The German offensive at Verdun in early 1916 witnessed the introduction of special fighter units which created a screen of Scouts behind which the German observation machines could operate in comparative safety. The French antidote was to use bombing machines, escorted by Scouts, to attack German ground troops. This forced the Germans to attack the bombers, leaving French observation aircraft to operate unmolested. The introduction of the Nieuport Scout fighter directly contributed to the initiative passing back into French hands in the skies over Verdun. On the British front, aerial superiority and protection for observation machines had been gained by July, for the beginning of the Somme offensive. This was mostly due to an administrative and tactical reorganization of the Royal Flying Corps carried out by Major-General H.M. Trenchard. The German response was to form units composed solely of fighter aircraft *(Jagdstaffeln)*. Boelcke was in command of Jasta II, and included amongst his pilots was Lieutenant Manfred von Richthofen. Reinforced with new aircraft such as the Halberstadt D.II, Fokker D III and Albatros D III, the balance had, once more, tipped in favour of the Germans by the end of the year.

By 1917, the essence of successful aerial combat was formation flying. The basic tactical unit had become the flight of six aircraft. Normally the flight consisted of the leader, in front, flanked above by two aircraft to form a V-shape; behind and above these were two more aircraft with the sub-leader bringing up the rear. The stepping up of the formation from front to rear gave maximum protection and allowed the flight to concentrate rapidly in the event of an attack. During combat, the aircraft usually operated in pairs, one to attack and the other to defend. From late 1917 onwards, both sides put larger patrols into the air. This was often done by layering flights of six aircraft, one above the other. However the problems of co-ordinating such large formations with the crude methods of communication then existing, created special problems which demanded the highest qualities of personal leadership.

The beginning of 1917 was marked by increased aerial activity, culminating in 'Bloody April' when British casualties were the heaviest for the whole war.

Manfred von Richthofen (in cockpit) and members of *Jagdgeschwader* 1. (IWM Neg Q42283)

German observation
balloon. (IWM Neg
Q41749)

This preliminary imbalance was partially redressed by mid-Summer with the introduction of the S.E. 5 and the Sopwith Camel. To counter the growing British numerical superiority, the Germans established large mobile fighter units *(Jagdgeschwader)*. These 'circuses' moved up and down the line according to the shifts in demand of local tactical situations. J.G. 1, commanded by Manfred von Richthofen, was formed at the end of June 1917.

The final year of the war was marked by German efforts to regain air superiority for the ground attacks of Spring 1918. The introduction of the Fokker D VII was an important influence on air operations in these months. However in August, all aerial activity became totally subordinated to the advance of Allied ground forces.

German prisoners
captured in an attack
on St Eloi craters in
March 1916. (IWM
Neg Q496)

Trench Raiding

From November 1914, trench raiding became the characteristic British method of dominating No-Man's-Land and securing information about enemy units. Some raids, organized at brigade and divisional level, were highly elaborate and took on the aspects of minor offensives. On the night of 14/15 September 1916, a series of raids on the Second Division front included a discharge of gas, field and heavy artillery bombardment, the blowing of mines and one minute of intense shelling by Stokes mortars. Involving hundreds of infantry in the actual raids, the object was to distract enemy attention from the main operations being conducted by II Corps further south.

The average trench raid usually had the limited objectives of killing or capturing enemy troops and destroying his dugouts, machine-gun posts and trench mortar positions. Personal equipment carried by the raiders was limited to rifle, bayonet, helmet and respirator. However many officers and men preferred to carry weighted clubs, daggers and knuckledusters for hand- to hand fighting. For mutual identification, armbands or pieces of ribbon tied to shoulder straps were worn. Torches were normally carried by officers and the searchers whose job was to find evidence from which the enemy units could be identified. All ranks had

British troops leaving their trench on a raid of the enemy lines near Arras on 11 April 1917. (IWM Neg Q5100)

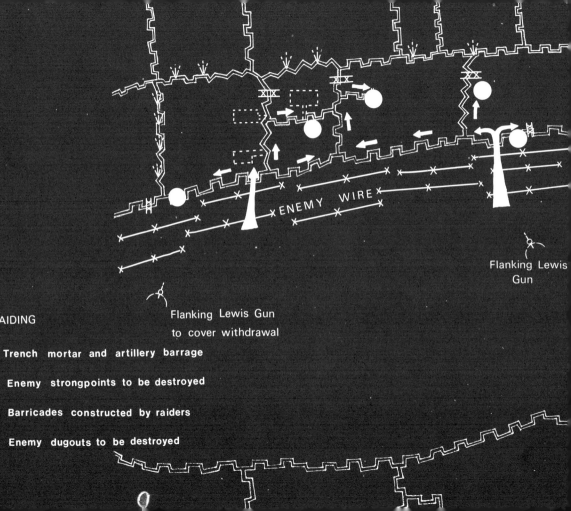

ENEMY WIRE

Flanking Lewis
Gun

Flanking Lewis Gun
to cover withdrawal

TRENCH RAIDING

Trench mortar and artillery barrage

Enemy strongpoints to be destroyed

Barricades constructed by raiders

Enemy dugouts to be destroyed

to leave maps, letters and other documents in their own trench before setting out.

The first obstacle, the enemy wire, would normally be cut by an artillery barrage or a wire-cutting party. The raiders would take up a position, close to the British wire, in No-Man's-Land at about five minutes before zero. At zero, an artillery and mortar barrage would be put down on the forward enemy positions. After about six minutes, the bombardment moved on to the second line trenches. This prevented enemy reserves from moving forward and protected the raiders' flanks. The raiders, supported by mortar and machine-gun fire, would enter the enemy line at two or three points. Once in the trenches, bombardiers would bomb from traverse to traverse, followed by the infantry. Lewis gun barricades would be established in flanking trenches to prevent counterattacks. A detachment of Royal Engineers usually carried out the task of destroying enemy dugouts and emplacements. After a specified period, the party would withdraw to their own lines, protected by Lewis gun postions in No-Man's-Land.

Trench Observation and Sniping

In trench warfare, troops rarely saw their enemy. However the policy of attrition demanded a steady toll of enemy casualties from artillery and small arms fire. For this reason, a combined system of small arms sniping and continuous observation became an important function of routine trench life. Each battalion normally had a special detachment of trained snipers operating under a selected Officer or NCO. Their basic duties were to keep the enemy's lines under constant surveillance, noting any new work carried out and the location of enemy strong points. At the same time, they were to keep the enemy's snipers in check and inflict casualties.

The sniping system adopted by the British Army consisted of a network of posts arranged along the front of each battalion sector. Each post observed a definite length of the enemy's front, throughout the hours of daylight. The snipers themselves normally worked in

Camouflaged observation post overlooking Lens in May 1918. (IWM Neg Q6628)

pairs with one observing through binoculars or periscope, whilst the other actually sniped at the enemy. Four men, in two reliefs, were told off to each post and they submitted a daily report on their activities and observations. When taking over a new section of line, the relieving snipers went into the line twenty-four hours before the battalion, so as to acquaint themselves with the front. In the German Army, snipers operated continuously in the same area. This familiarity with a single section of front certainly contributed to greater efficiency.

No definite rules were laid down for the location of sniping posts. This was usually left to the skill and ingenuity of the individual. Positions could be found in shell craters in No-Man's-Land or in elevated posts behind the front line. In the fire trenches, steel loopholes were usually built into the parapet to provide protection for the snipers and observers. The concealment of loopholes and observation posts was helped by making the outer faces and top of the parapet irregular. The Germans broke up the line of their parapets by piling

Snipers wearing camouflage suits. (IWM Neg Q65492)

timber beams, bolsters, mattresses and other rubbish against them. However even strong well-camouflaged positions were vulnerable to the artillery, which was often called upon to deal with a troublesome sniper's post.

A German
observation tree
captured by the
Canadians near Arras
in 1918. (IWM Neg
C.O. 1973)

An Australian officer
wading through the
mud of 'Grid Trench'
in December 1916.
(IWM Neg E572)

12   1917: The Allies at Bay

In March 1917, the Germans upset French offensive plans by retiring to the solidly constructed defences of the Hindenburg Line. Undeterred, the new French Commander, General Robert Nivelle, launched two Spring offensives at Arras and on the Aisne to test his vaunted formula for tactical success. By May, the attacks had failed and the French Army was in a state of mutiny. The situation was only restored by General Pétain who placed the French on the defensive for the rest of the year. Meanwhile the British attempted to break the German flank by a direct attack in the Ypres area. Messines Ridge was taken in June and, in July, the Third Battle of Ypres began. The opening bombardment wrecked the drainage system of the area which,

combined with a wet summer, bogged the attack down. When the offensive died on 10 November, 245,000 British casualties had paid for a muddy topographical feature known as the Passchendaele Ridge.

In May 1915, the Italians had joined the war on the Allied side and, by 1917, found themselves involved in a war of attrition with which their resources could not cope. Heavily subsidized by Britain, they failed to drive the Austrians out of their mountain positions on the Isonzo river. In October 1917, the Germans sent reinforcements to their Austrian allies for a limited offensive launched at Caporetto on 24 October. The attack turned into a rout of the disorganized Italian Army. Seventy miles and 300,000 casualties later, the

Dead German sniper.
(IWM Neg E2955)

Behind the lines: Generals Joffre, Haig and Foch at Beauquesne in August 1916. (IWM Neg Q951)

Italians established a shorter line on the Piave which was successfully held for the rest of the war.

The most serious blow to the Allies during 1917 was the elimination of Russia from the conflict. In 1916, General Alexei Brusilov's offensive had been spectacularly successful, but was bought with 1,000,000 Russian casualties. This became another contributory factor in the worsening internal situation. In March 1917, the Tsar abdicated and, in the following July, the Provisional Government threw the crumbling army into another offensive in Galicia. By the end of the year, Russian military strength had melted away and the Bolsheviks had seized political power. Although the dissolution of the Allies' Eastern Front was an important factor, it was more than compensated by America's entry into the war, on the Allied side, in April 1917. However it would be mid-1918 before America's weight was felt on the Western Front.

13 Discipline and Morale

Conditions Behind the Lines

Following their period of front-line duty, the troops were relieved and moved into reserve billets. These billets were usually abandoned farms and buildings, lacking any domestic comforts and usually within range of enemy artillery. However fresh meat and, less often, fresh vegetables were available for the men. On the first day in reserve billets, the only military duty was rifle inspection with the rest of the day free for the troops to wash themselves and clean their uniforms. On the second morning, Company Officers' inspection took place. For this parade, the men had to be clean and shaven, with their uniforms and equipment in good order. Following the inspection, there was usually one hour of rifle practice or physical exercise after which the troops

were dismissed. Although in reserve positions, units could still be called out at any time to man defensive posts or be loaned to the Engineers for reserve trench digging. The several days spent in reserve billets were, therefore, a mixture of rest and readiness with the advantages of regular mail and a wash-tub.

After two weeks in trenches and reserve billets, the brigade moved into rest camps for about a week. The first stop made by a relieved unit was generally at the bath-house or de-lousing station. This establishment was usually situated in a disused farm, brewery or laundry and was equipped with large wooden tubs. Hot water, heavily laced with disinfectant, was available and the men also received fresh shirts and underclothing. In rest camps, there was

a regular distribution of mail and presents of cigarettes, tobacco, socks and other woollens sent from home. In addition to training in musketry, tactics, mortaring, bombing and the inevitable parades, time was also available for sporting activities. Variety programmes and boxing were often arranged to provide evening entertainment for the troops.

For the Non-Commissioned Ranks, home leave was comparatively rare and was impossible for everyone during the months preceding a big offensive. For many of the troops, the only way to get home was to receive a 'Blighty one'. This was a wound sufficiently serious to take the recipient back to England. That men were willing to loose a limb in exchange for a permanent return home, reflected their war-weariness.

Mutinies and Courts-Martial

It is a paradox of military discipline that, ultimately, it is only maintained through the willing acceptance of the Enlisted Men and the Non-Commissioned Officers. The Russian Revolution illustrated the powerlessness of

Officers and policing authorities to cope with problems of mass mutiny. During the summer of 1917, the French Army faced a similar problem when mutiny became rife among front line units. The outbreak coincided with the evaporation of the Nivelle offensive into a series of despairing attacks to secure a number of geographical features along the Chemin des Dames. Badly mauled regiments were pulled out of the trenches for a short rest and then fed back into the offensive. From the beginning of May, troops refused to return to the front and, during the month, there was an accumulation of unrelated incidents. The situation was aggravated by pacifist and left-wing agitation and the presence, in the French line, of two Russian brigades that were teetering on the edge of open revolution. By June, the troops were criticizing conditions behind the lines, particularly the rest camps and lack of leave, as well as the conduct of the war by the French High Command and Government. In early June, it was estimated that only two divisions could be wholly relied upon and the front was being manned with

British troops receiving dinner rations from field kitchens. (IWM Neg Q1582)

the remnants of once strong divisions.

The spontaneity of the mutinies and the numerous motives for the unrelated outbreaks prevented the creation of an integrated revolutionary force. On 15 May, General Henri Pétain replaced Nivelle. Conditions for the troops were gradually improved and offensives limited to well prepared tidying-up operations. At the same time, Pétain showed no compunction in executing leading mutineers. By the end of 1917, the French Army was capable of holding its front line but could not be risked in an offensive.

In the British Army, the problems of widespread mutiny never had to be faced. Officers and military police, supported by the influential Non-Commissioned Officers, maintained routine discipline within the expanding force. However, according to official figures, 332 British soldiers were executed in France and Belgium for a variety of military crimes. The majority of the sentences were for desertion, followed by cowardice and murder. Five men were executed for disobedience and two for sleeping at their posts.

Honours and Awards

Orders, decorations and medals have been distributed for military service since the eighteenth century. They provided an incentive for courageous conduct and good service, as well as indicating the experience of an individual or regiment. During the First World

The town of Albert on the Somme: It was said that when the leaning Virgin fell, the war would end. (IWM Neg C.0.2132)

120

War, both sides used them to encourage troops and boost *ésprit de corps*. The Allies tended to be more liberal in the institution and distribution of awards.

Within the British Army the highest award for gallantry was the Victoria Cross, which had been instituted in 1856 by Queen Victoria. This decoration, which took precedence over the British Orders of Knighthood, was available to both Officers and Men and could be awarded posthumously. From August 1914 until the end of the war, a total of 633 VCs were distributed, 517 being won in France and Belgium. The German equivalent was the *Ordre Pour Le Merite* or 'Blue Max'. Other British gallantry awards included the Military Cross for Commissioned Officers, of which 31,793 were distributed; the Distinguished Conduct Medal for NCOs and Men, 21,041 being awarded on the Western Front; and the Military Medal also for NCOs and Men, of which 110,342 were awarded.

However most British troops had to be satisfied with one of the two campaign medals in addition to the British War Medal (1914–18) and the Victory Medal. These were issued in recognition of service in a theatre of war during a specified period. The Orders of Knighthood were reserved for the senior commanders. The most common award to generals was one of the three classes of the Military Division of the Most Honourable Order of the Bath. Sir Douglas-Haig, Commander-in-Chief of the B.E.F. from 1915, was granted an earldom and £100,000 for his services. The relatives of soldiers who had been killed during the war received a 'Next of Kin Memorial Plaque' and a commemorative scroll from the King.

14   1918: The Re-instatement of Mobile Warfare

The Strategic Considerations

By the end of 1917, Germany was beginning to experience the military and economic effect of the British naval blockade. The German export trade had been virtually extinguished and the German war industry was being denied essential imports of raw materials. Food rationing had been introduced and now added to an accumulating picture of eroding public morale. A German effort to strangle Britain's economic life through the reintroduction of U-boat warfare had failed. The use, by Britain, of the convoy system, improved mines, and the increased activity of British submarines had countered the U-boat campaign. This second German submarine offensive, however, was a major contributory factor in the decision taken by the United States to declare war on Germany in April 1917.

The entry of America into the war added a new dimension, in terms of the balance of economic and military forces, to the fighting on the Western Front. Until 1917, the trench deadlock had been maintained by the industrial and military parity of the combatants. However American power was of such a scale that it dwarfed all the resources of any combination of the existing sides. If Germany sought a decisive military victory which would end the war, then it would have to be achieved before America could deploy her forces on the Western Front.

The dissolution of Russian military power during 1917 effectively released nearly a million men for use in a German offensive on the Western

Troops drinking coffee supplied by the Y.M.C.A. near Ypres in 1918. (IWM Neg Q8397)

Front. This injection of manpower meant that a temporary numerical imbalance could be created in the areas selected for attack. On the other hand, Great Britain had only been able to maintain her forces through the increasing commitment of Empire troops to flesh out the skeletal remains of the volunteer armies.

Committed to final victory on the Western Front, Germany was obliged to strike the Allies in Spring, 1918. The choice of the Somme area for the attack was based upon considerations of terrain and the strategic decision to unhinge the joint between the British and French and then smash the British Army in detail. These cloudy, open-ended strategic objectives failed to match the tactical originality of the planned offensive.

The Canadians take Vimy Ridge in April 1917. (IWM Neg C.O.1162)

American bombardiers moving forward. (IWM Neg Q69942)

Germany: New Tactics and the Spring Offensives

In 1918, the trench system still represented a formidable barrier which had yet to be successfully breached by an attacking force. Both sides had adopted the defensive methods introduced by the Germans in 1917. The continuous front line and support trenches of 1915 and 1916 had been replaced by an outer line of heavily wired-in and camouflaged strong points. These were backed by a deep defensive Forward Zone, consisting of mutually supporting machine-gun positions, swathes of barbed-wire and open areas which could be bombarded by the defender's artillery. Several miles behind this line was the main Battle Zone, incorporating strong points, deep concrete dugouts and trenches. The third defensive area was the Rear Zone, situated four to eight miles behind the Battle Zone, in which the majority of the defending troops would, in theory, be held. In March 1918, the British system was still under construction and, in the Fifth Army area, only the Forward Zone had been completed. Secondly,

British defensive tactics had failed to adapt to this flexible system, relying instead upon the linear doctrines of 1915–16.

Committed to unconditional victory on the Western Front, the Germans made a thorough-going search to find the tactical key that would unlock the Allied defensive system. The tactics and organization which evolved were based upon the use of specially trained units of *Sturmtruppen* (storm troops). These consisted of the youngest, most experienced and physically fit men that could be siphoned off from the existing regiments. Their task was to cross the enemy lines, by-passing strong points and areas of resistance, to break through the defensive zones and strike the enemy artillery and reserves. The first rush of the spearhead was followed by the rest of the *Sturmbataillonen*, consisting of four pioneer companies, machine-gun section, flame-thrower and trench mortar detachments. The role of these troops was to support the initial assault waves and eliminate enemy pockets of resistance. Finally the conventional infantry units moved in to

Hindenburg (front left) and Ludendorff (front right) with Ukrainan visitors on 9 September 1918. (IWM Neg Q45363)

mop up the remaining enemy troops. Ideally, reserves were only fed in to support a successful assault, thereby increasing the attack's momentum, widening its front and deepening the penetration. These fluid tactics were made possible by a highly flexible system of command which placed wide tactical decisions, usually reserved for General Officers, at the discretion and initiative of Officers in the field.

Shortly before dawn on 21 March 1918, the Germans opened their offensive with a brief hurricane bombardment by 6,000 artillery pieces of all calibres. Working to a fire-plan designed by Colonel Bruchmüller of the Eighteenth Army, the guns opened initially upon enemy batteries, communications, strong points and command posts. Surprise was increased because the guns had been laid by means of survey and information supplied by air reconnaissance, instead of the telltale system of direct registration on the target. Under the surprise and concentration of the bombardment, British communications with forward units gradually disintegrated. For five hours, the German guns moved systematically from target to target, pounding vital defensive locations. Mustard gas was used extensively against strong points which were likely to threaten the flanks of the assault. The bombardment culminated in a final crescendo of

German Stormtroops practising in the use of smoke at the Infantry Training School at Sedan. (IWM Neg Q53091)

Men of the 4th Canadian Division at the Canal du Nord on 27 September 1918. (IWM Neg Q9326)

high explosive shell on the British forward position. The *Sturmbataillonen* attacked under cover of fog and achieved a breakthrough into open country along a forty mile front. Within a week, the Germans had advanced thirty miles and were within reach of Amiens where they threatened to sever communications between the British and French armies. Yet, by the end of the month, they had failed to capture Amiens and the Michael offensive had slowed and shuddered to a halt.

Although the German High Command had partially solved the tactical problems of the offensive, it had ignored the logistical lessons of 1914. Lacking sufficient motor transport, supplies of food and ammunition, they could not sufficient motor transport, supplies of food and ammunition could not keep pace with the advance. This problem was accentuated by the added difficulties of provisioning troops across the desolate wastes of earlier battlefields. Thus heavy casualties and exhaustion crippled the offensive momentum that should have dominated

129

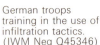

**GERMAN
INFILTRATION
TACTICS**
The units of
Sturmtruppen crossed
the enemy lines, by-
passing strong points
and areas of
resistance, to break
through the defensive
zones and strike the
enemy artillery and
reserves. This
spearhead was
followed by the
rest of the
Sturmbataillonen,
consisting of
Pioneers, a machine-
gun section,
flamethrowers and a
trench mortar
detachment.

the Allies and prevented them from any co-ordinated response.

Subsequent weaker offensives, from April to July, in the Ypres-Bethune and Soissons-Reims areas took the Germans briefly across the Marne and to within forty miles of Paris for the second time during the war. However the Allies managed to contain these efforts on all fronts. By mid-July, the Germans were still facing an unbroken enemy but this time along an improvised line, held by the exhausted remnants of their assault forces. Moreover the deployment of American troops and the integrated use of tanks by the Allies on the Western Front, was about to create a tactical imbalance which Germany could not effectively redress.

The Allies: Tanks and the Final Victory
The introduction of the tank by the

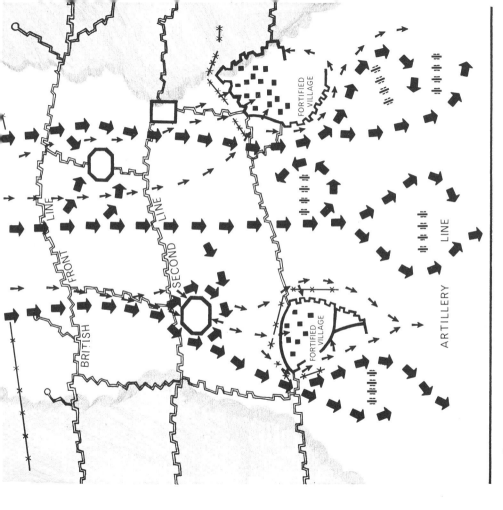

German Infiltration Tactics

Initial wave of stormtroops

Heavily armed support troops and infantry

Mustard gas covering flanks of assault waves

Strong-points and heavily defended localities

Artillery

Allies represented the first real application of mechanized power to solve the tactical problems posed by a war between comparable industrial powers. The German offensives of Spring 1918, had illustrated that human muscle and endurance alone could not win in a war of steel. Thus the mobility and armour of the tank which enabled it to cross barbed wire and trenches and to penetrate strong defensive zones, presented an offensive antidote to defensive strength.

The original idea for a multi-purpose offensive machine was conceived by Lieutenant Colonel (later Major General Sir) Ernest Swinton. It was also due to his persistency, in spite of high-level opposition, that the tank was introduced into service. On 2 February 1916, official trials were conducted at Hatfield and 190 of the machines were subsequently ordered. In the following September, they were used for the first time during the Somme offensive. However, owing to tactical mismanagement and technical defects, the results were disappointing. One year later, at Cambrai in November 1917, tanks gave a brief but emphatic display of their tactical potential. Without any preliminary bombardment of enemy positions, nearly 400 tanks attacked and achieved total surprise over the enemy. An unprecedented initial advance of five miles for 1,500 casualties was gained, only to be lost because there were

German transport column on the Albert-Bapaume road in March 1918. The problems of supplying a rapid advance contributed to the deceleration of the German offensives in 1918. (IWM Neg Q60474)

no reserves available to maintain and exploit the unexpected success. The German counter-attack which recaptured all the lost ground relied upon the infiltration tactics that were to be so effective in 1918. Tanks were also used in a number of the minor operations which preceded the main Allied attacks in July and August 1918. At Le Hamel, on 4 July, a brief bombardment and creeping barrage, together with the strafing and bombing of enemy positions by aircraft, were carried out in conjunction with a rapid advance by a concentrated force of tanks. This small exercise set the tone for the British offensive launched near Amiens in August.

The preliminary move in the Allied counter-attack had been made by the

A tank dealing with barbed-wire at Wailly in October 1918. (IWM Neg Q6425)

A German tank being de-trained in July 1918. During the war, the Germans failed to recognise the potential of the tank. (IWM Neg Q55377)

French on 18 July. Eighteen divisions, together with 225 tanks, had struck the improvised German salient which had been created by the earlier German offensive on the Marne. The second stage was the Amiens offensive mounted by the British on 8 August. For this attack the Fourth Army had gathered thirteen divisions, three cavalry divisions, 2070 guns, 324 heavy tanks, ninety-six light 'Whippet' tanks and 120 supply tanks. A large number of aircraft were also made available for attacks on enemy ground positions. The actual attack was made through a morning mist, behind a creeping barrage; the infantry advancing behind the tanks with the reserves in close support. The tanks and infantry succeeded in breaking the German line and created an eleven mile gap.

However heavy tank losses and failure of attempts to co-ordinate cavalry with tanks in a battle of exploitation, combined to rob the British of a decisive victory. Ludendorff called 8 August 'the black day of the German Army', but his troops were not broken and were still able to maintain a dogged and orderly retreat. By the beginning of September, the Germans had been remorselessly pushed back into the Hindenburg Line. The Allied Front had acquired a cohesive momentum as it tumbled across France with the Germans retreating before it. The impetus for the advance had been created by the tank which had unlocked the trench barrier and forced the fighting out into the open. In spite of heavy losses in tanks, the initial velocity of the attack did not die as it had done at Cambrai in 1917. This was largely due to the Allies' overwhelming reserves in manpower and *matériel* which gave no respite to the enfeebled German forces. At the end of September, the Hindenburg Line was breached after only two days of assault and on 4 October Germany requested an armistice. While the terms were being discussed, the bitter fighting continued on the Western Front. However Germany had already disintegrated, both politically and economically, which made it difficult for her to sustain prolonged and heavy fighting. Active hostilities finally ceased on 11 November with the Armistice.

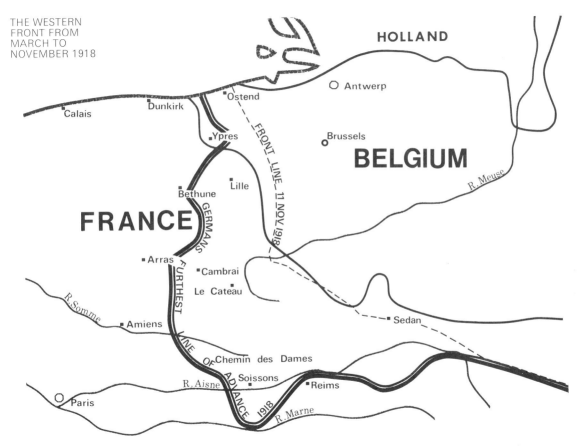

THE WESTERN
FRONT FROM
MARCH TO
NOVEMBER 1918

HOLLAND

○ Antwerp

■ Ostend

Calais

■ Dunkirk

■ Ypres

FRONT LINE 11 NOV 1918

○ Brussels

BELGIUM

R. Meuse

■ Lille

■ Bethune

FRANCE

GERMANS

■ Arras

FURTHEST

■ Cambrai

■ Le Cateau

R. Somme

■ Amiens

LINE OF ADVANCE

Chemin des Dames

■ Sedan

R. Aisne

■ Soissons

■ Reims

○ Paris

1918

R. Marne

15 The Cost

During four years of continuous warfare, the combatants had poured out their human, economic and physical resources for no other ostensible reason than the need to defeat the enemy. Great Britain and the Empire had mobilized 9,000,000 men of which nearly one million had been killed and over two million wounded. Total casualties to France were nearly twice those of Britain in killed and wounded. Germany had mobilized 11,000,000 men of which nearly six million had been either killed or wounded. For the three nations, these figures represented the virtual loss of the rising generation. In addition, the fighting had absorbed about ten years of normal economic effort in just fifty-one months. Both Great Britain and Germany each spent about £12,000,000,000 in financing themselves and subsidizing their allies. The United States loaned $9,452,000,000 to the Allies throughout the war.

The subsequent peace settlement worked out at Versailles gave little or no indication of why the human and economic resources had been squandered. The primary aim of the three major powers was to make Germany incapable of further military aggression and to settle Europe under the supervision of the League of Nations. The Armistice terms had already begun the military emasculation of Germany. 5,000 artillery pieces, nearly 25,000 machine-guns, 1,700 planes and most of the German Navy, including all submarines, had been surrendered under its terms. In 1919, the Allies tried to solve the problem of Germany's potential power by imposing firm restrictions on her armament growth. Yet

Tanks moving through Meaulte shortly after its capture in August 1918. (IWM Neg Q7302)

139

the fact that France retained a large standing army indicated that Allied superiority in Europe would only be temporary. This situation was accentuated by America's refusal to participate in the League of Nations and her withdrawal from involvement in European political affairs. This also served to compromise the international viability of the League since members preferred to ignore it when war suited their political objectives.

During the immediate post-war period, the belief that the appalling human and economic cost of the war would preclude the possibility of future conflicts impaired the military thought of Britain. Relying upon a policy of diplomatic appeasement, Britain sacrificed her military preparedness. The tactical lessons of mechanization and the influence of armoured forces were ignored. In France also, there was a total miscalculation in terms of military planning. Stress was now placed upon the superior strength of the defensive, ignoring the offensive factors that had made modern warfare a struggle between highly mobile and flexible forces. In contrast, the Germans coupled infiltration methods with mechanization and air power to formulate the tactics of *Blitzkrieg* that were to be so successful in Europe from 1939 to 1942.

The tactical developments of trench fighting guaranteed that large unsupported infantry forces would never again be pitted against each other in a war of static, remorseless attrition. However the evolution of tactical expedients in no way justified the human and economic sacrifices made by the combatants on the Western Front between 1914 and 1918.

German soldiers killed by shellfire. Over 53% of Germany's mobilized forces were either killed, wounded or taken prisoner during the war. (IWM Neg Q5733)

A Select Bibliography

R.H. Beadon · *The Royal Army Service Corps, Vol 2.* · Cambridge University Press

Correlli Barnet · *The Swordbearers* · Eyre and Spottiswoode

J.E. Edmonds · *Official History Of The War: Military Operations.* · 27 Volumes Macmillan & Co Ltd and H.M.S.O.

A. Farrar-Hockley · *Death of an Army* · Arthur Barker Ltd

B.H. Liddell Hart · *History of the First World War* · Various editions

B.H. Liddell Hart · *The Tanks, 1914–45, Vol I* · Cassell & Co Ltd

History of the Corps of Royal Engineers Vol V · The Institution of Royal Engineers.

A. Horne · *The Price of Glory* · Macmillan & Co Ltd

R.F. Nalder · *The Royal Corps of Signals* · Royal Signals Institution.

Statistics of the Military Effort of the British Empire during the Great War, 1914–1918 · H.M.S.O.

B. Tuchman · *August 1914* · Constable & Co Ltd

E. Wyrall · *History of the Second Division, 1914–18* · Thomas Nelson & Sons Ltd

Index